Chiral Reactions in
Heterogeneous Catalysis

Chiral Reactions in Heterogeneous Catalysis

Edited by

Georges Jannes and Vincent Dubois

CERIA / Meurice Institute
Brussels, Belgium

Plenum Press • New York and London

CHEM

Library of Congress Cataloging-in-Publication Data

On file

Proceedings of the First European Symposium on Chiral Reactions in Heterogeneous Catalysis, held October 25–26, 1993, in Brussels, Belgium

ISBN 0-306-45110-7

© 1995 Plenum Press, New York
A Division of Plenum Publishing Corporation
233 Spring Street, New York, N. Y. 10013

10 9 8 7 6 5 4 3 2 1

ORGANIZING COMMITTEE

A. Baukens
V. Dubois
G. Jannes
J.P. Puttemans
P. Vanderwegen

SCIENTIFIC COMMITTEE OF THE CENTENARY

Prof J. Vandegans, chairman
Prof. A. Blondeel, treasurer
Prof. A. Debourg
Prof. P. Dysseler
Prof. G. Jannes
Prof. J. Hanuise
Prof. T. Lepoint
Prof. C.A. Masschelein
Dr. J.P. Simon

PREFACE

It was a great honor for us to organize ChiCat, a symposium devoted to Chiral Reactions in Heterogeneous Catalysis and to be the hosts of more than 120 scientists coming from everywhere in the industrialized world, to celebrate together one century of existence of Institut Meurice.

This school was established in 1892 when an industrial chemist, named Albert Meurice, decided to educate practical chemists according to the perceived needs of the industry of that time. This is exactly what we are still trying to do. It is the reason why, thirty years ago, we started a research activity in catalysis, and why we progressively devote this research to the applications of catalysis in the field of fine chemicals. In this respect, we are very close to another initiative of Albert Meurice, who started the first production of synthetic pharmaceuticals in Belgium during World War I. This business later on became a part of the Belgian corporation UCB, still very active in pharmaceuticals today.

The school created by Albert Meurice merged in the fifties with another school that had been created to meet the same needs in the field of the food industries, mainly distilleries and breweries. This merger was done in the frame of the establishment of CERIA. For people in catalysis, ceria stands for cerium oxide, but for those who engineered the concept, CERIA stood for Center of Education and Research for the Food and Chemical Industries. Our center was created to stimulate the rebuilding of these industries in the province of Brabant, ruined by the war. Today it plays a prominent role in the development of technopoles in the new region of Brussels organized in federal Belgium. This explains the financial support we received from Brussels Ministry of Economy for the Centenary of Institut Meurice.

However, this symposium is probably the last meeting of the old fashioned Contact Group – Catalysis as supported by the National Science Foundation. The structure of financial support to scientific research has also been modified. We hope that it will not result in further diminishing the contribution of the national and local governments to the necessary endeavor of Belgium in research and development.

Twenty years ago, the Société Royale de Chimie founded its Catalysis Division. The first president of this Division, Professor Delmon, accepted that the first event organized by the division should help celebrate the twenty–fifth anniversary of CERIA : it was the First International Symposium on the Relations between Homogeneous and Heterogeneous Catalytic Phenomena, a series that is still going on. Nowadays, for the centenary of Institut Meurice, we are pleased to organize this meeting in the frame of the Catalysis Division of the Société Royale de Chimie. It is good to have some traditional friends when celebrating anniversaries.

A newcomer in the Belgian Catalysis community is the Catalysis Committee of the Royal Belgian Academy Council of Applied Sciences. This Committee has issued a report on Modern Catalysis that has received positive echoes in Japan as well as in the United States : we hope that it will not remain without answer in Belgium. The Royal Academy Council has accepted to give its patronage to our symposium ; it is a great honor, and we are thankful for this.

Finally, I would like to acknowledge the fantastic work done by my co–workers and to publicly thank them. I am proud to have them as co–workers and friends. Vincent Dubois did a huge part of the long term organization, sacrificing more than six months of his Ph.D. research to prepare this event. Anne Baukens, Danielle Philip, Maryse Talbot, Pascal Vanderwegen and Jean–Pierre Puttemans also devoted time and energy to make attendance as effective and pleasant as possible. Thanks are finally due to the staff of our Institute, to the Scientific Committee of the Centenary, and to some colleagues in our Center, especially in the hotel school.

Georges Jannes

ACKNOWLEDGMENTS

The following companies agreed to provide financial support to the symposium :

        Amer-Sil
        Catalysts and Chemicals Europe
        Fina Research
        Hewlett-Packard
        Janssen Pharmaceutica
        Johnson-Matthey
        Sanofi Pharma
        Solvay
        UCB
        van Lerberghe

The symposium was also supported by :

        FNRS-NFWO (Belgium's National Science Foundations)
        Ministère de l'Economie de la Région de Bruxelles-Capitale

The following companies participated in the permanent exhibition :

        Analis (Parr)
        Autoclave Engineers
        Janssen Chimica
        Micromeritics
        Sigma-Aldrich
        Technology Catalysts

The organizers are grateful to them for their generosity.

CONTENTS

HYDROGENATION SYSTEMS : BROADENING THE SCOPE

MORE ADVANCED REACTIONS

Chiral Reactions in
Heterogeneous Catalysis

INTRODUCTION

During the last decade, demand for enantioselectivity has grown, and increasing force and funding have been directed toward preparing pure chiral compounds.

Catalysis, especially heterogeneous catalysis, is probably the most desirable means of attaining that objective, but it has still to demonstrate its full ability to reach that aim.

It was a daring enterprise to launch ChiCat, a symposium on chiral heterogeneous catalysis. The topic was, and still is, very mobile, and controversial. That is the reason why we decided to also invite lecturers from outside the field of heterogeneous catalysis, to help us to better understand the general context.

Dr. Polastro put the question in a market perspective: he opened the symposium with a revealing look on the future market of the chiral chemicals we are striving to prepare in optically pure states. Professor Reisse warned the catalysis people not to reinvent the wheels that organic chemists have labored over many years. Professor Ghosez challenged the attendees by showing that the enantiomeric excesses that organic synthesis is able to achieve set up ultimate goals that will not be easy to attain by heterogeneous catalysis, and that, moreover, enantioselective heterogeneous catalysis is not only limited today in its performances, but also in its capabilities: (little C — C bond formation, if any...). Homogeneous catalysis is offering performances and concepts that are very stimulating for our future developments: they have been outlined by Professor Brunner. Different strategies are developed to get enantioselectivity in heterogeneous catalysis: use of a chiral modificator on the surface, grafting of a chiral catalytic complex on a solid (the so-called heterogenized homogeneous catalysis), and homogeneous formation of a chiral complex before surface reaction. We invited pioneers in the field to detail these strategies: Professor Webb, Professor Pini, Dr. Blaser and Professor Tungler. We thank them all for their kind acceptance and excellent and stimulating lectures. We also appreciate the quality of the contributed posters and recognize the special effort the authors made in trying to fit as precisely as possible the topics of the symposium. A special session was devoted to short presentation and discussion of these posters, giving the audience an overview of the last developments of chiral catalysis, in terms of new catalytic systems and new substrates as well. We are very grateful to the chairpersons for their kind authority in maintaining a correct timetable.

The members of the Scientific Committee of ChiCat helped us to build a program with the best contributors we could hope to have. They were faced with the difficult task of keeping the program focused. They are all most sincerely thanked for the outstanding job that they accomplished.

The symposium delineated what is really at stake. And different routes are paved to attain the goal. But are we sure that those routes will go so far?

Hydrogenation reactions remain the mainspring of the field, with alkanones and $\alpha-$ or $\beta-$ketoesters as preferred substrates, and nickel-tartatic acid and platinum-cinchona as preferred catalytic systems. Extrapolation to other substrates: aromatic ketones, acids, and ethers, steroid ketones, Schiff bases,... has been attempted, with increasing success. On the other hand, substantial work has been devoted to the discovery and the interpretation of the effective parameters: nature of the metal, effect of the modificator, place where the enantio-determining step arises, hydrogen pressure, influence of the support. Knowledge gained in this way has been applied to vicinal reactions: dihydroxylation, for instance, and to reactions that are considered as more challenging, at least in heterogeneous systems: dehydration epoxidation, and cyclopropanation.

Very fine catalyst preparations, using for instance organometallics and zeolites as building blocks, and inclusion methods[1] were also described.

Perhaps, the very classical approach should be tried again: looking at the global problem. Chemical engineering studies, namely of the mass transfer in the reactor, could contribute to the definition of the exact amount of co-reactant, hydrogen for instance, that is needed at the surface of the catalyst. Anyway, this will be necessary if we want to be able to scale up our laboratory achievements. Solution chemistry will help us to better understand the way a prochiral molecule may be complexed before its adsorption, but also to discover some side effects of the modification process, metal leaching, e.g., and their effects on the process. Surface science sheds some light on the manner in which the catalyst surface can be chemically modified to build a chiral site. Adsorption studies will lead to a better picture of the modification process and of the activation process as well. As mentioned, the choice of special supports and the preparation of engineered supports also provide possible ways to reach this goal.

During the symposium, we had an informal dinner to discuss the future of ChiCat: the decision was taken to give a follow-up to the symposium. We strongly believe that we need another place to meet than in the frame of the broader Chirality meetings, or in big Heterogeneous Catalysis Congresses. Instead of starting a new series of symposia, the participants to the discussion preferred the formation of a ChiCat Group, inside which we will try to find the best way to interact, to cooperate, and to exchange ideas, in a more informal way. This would also be the place to stimulate young scientists' mobility. This group could meet as a satellite to well-chosen symposia, of vicinal but broader topics. The next symposium on Heterogeneous Catalysis and Fine Chemicals, which will be organized in Switzerland by Dr. Hans Blaser could be a good instance of the opportunities we are looking for. The idea is open to all those who are interested. What ChiCat started as a symposium, ChiCat group will take over!

V. Dubois
G. Jannes

---

[1] This contribution, by R. Selke, P. Bathelemy and J.P. Roque, is not included in the Proceedings.

SECTION I

THE PLACE OF HETEROGENEOUS CATALYSIS

IN THE CHIRALITY FIELD

COMMERCIAL OUTLOOK FOR CHIRALITY. QUO VADIS ?

E. Polastro

Arthur D. Little
Bd. de la Woluwe 2
B–1150 Brussels, Belgium

ABSTRACT

Over the past few years the field of chirality (defined as chiral synthesis and chiral fine chemicals) has attracted considerable interest from scientists, industrialists and more recently also investors.

It is interesting to see emerging similar patterns between this recent interest in chiral synthesis and the developments the financial, scientific and industrial community has experienced in the late 70's – early 80's with the advance of genetic engineering and modern biotechnology.

Such similarities are somewhat puzzling, given the numerous successes but even more numerous failure stories of ventures in the field of modern biotechnology. A more focused approach and more realistic hopes could have been expected for ten years later in the field of chirality.

Indeed, while the future development potential offered by chirality is highly promising, several key questions remain to be addressed, such as :

- Who will ultimately cash in from development in this field ?
- Where lies the value added ?
- What level of integration will be required ?
- ....

Probably an in–depth objective analysis would yield to a major reassessment of the strategies followed by many players in the field of chirality.

A FIELD ATTRACTING CONSIDERABLE INTEREST

Since several years already, it is almost impossible to find a single issue of a journal dealing with fine chemicals or organic synthesis that does not contain at least an article on

Table 1. A brief dictionary of stereochemistry (Source : Arthur D. Little).

Chiral	Describes a compound having one or more centers of asymmetry (usually a carbon atom). The asymmetry centre permits the compound to exist in two or more stereospecific configurations, or stereoisomers. (Derived from the Greek word *chiros*, meaning "hand", the term "chiral" suggests "handedness" and is applied to molecules having structures that are mirror images of each other, analogous to the left and right hands.)
Stereoisomers	Species of compound that differ in their spatial configuration, the species share the same molecular formula and most of their chemical properties. However, their physical properties may differ.
Enantiomers	Stereoisomers that are mirror images of one another. Depending on the number of asymmetric centers, a compound can have several pairs of enantiomers, each one of the pair called an enatiomorph.
Diastereoisomers	Stereoisomers that are mirror images of one another. In compounds having more than one pair of enantiomers, enantiomorphs in one enantiomeric pair will usually be diastereomers of the enantiomorphs in another enantiomeric pair.
Optical activity	An indirect measurement of a compound's chirality. The asymmetry in a chiral compound affects the passage of polarized light through a solution of the compound. Each of the possible stereoisomers affects that passage of light somewhat differently. The differences can be detected using a specialized polarimeter that measures the extent to which a dissolved sample of the compound rotates a plane of polarized light passing through it.

Source : Arthur D. Little

Table 1. *(continued)*

Optically active	Describes a compound that can exist in stereoisomeric forms. Same as c chiral.
Optical purity	The extent to which polarized light is rotated by only one stereoisomer of a compound, thereby indicating the stereoisomeric homogeneity of the sample. A completely heterogenous mixture of the possible stereoisomers will exhibit zero net rotation. A completely homogeneous preparation will exhibit the rotation characteristic of only one of the stereoisoemrs.
Optical pure	Describes a homogeneous preparation consisting of only one stereoisomeric form of a chiral compound.
Homochiral	Same as optically pure.
Enantiomerically pure	Describes a preparation of a compound that contains only one enantiomer. Such preparations are also described as optically pure or homochiral.
Racemic	Describes a preparation of a compound containing equal proportions of enantiomers.
Biologically active	Describes a compound that elicits an intentional biologic response, typically by binding to a cellular receptor. An intentionally toxic compound can also be described as biologically active, depending on its mode of action. Typically, the biological activities of the various stereoisomers are different.

Source : Arthur D. Little

the subject of chirality (see for a few definitions table 1). Similarly, the number of patents and symposia focusing on this topic is daunting.

Some recent surveys have suggested that more than 60% of all patent and scientific literature in organic chemistry is devoted to "chirality–related subjects" with *chiral* appearing as a key word.

This increasing interest of the academic and of the industrial community is reflected by several universities and chemical companies devoting significant efforts to develop their capabilities in chirality. Also, some start–ups have been created focusing specifically on this field.

It is therefore not surprising that the financial community is now discovering the field of chirality, identified as a possible source of new stock offerings and other forms of fund raising for technology driven start–ups.

It would be pointless and certainly beyond the scope of this paper to have the pretension of providing the ultimate answer on whether such a high level of interest in chirality is warranted and justified in terms of possible reward potential.

Indeed, while significant developments are expected in the field of chiral fine chemicals, considerable debate still exists on :

- the precise timing for the expected take–off of market demand, the issue being not whether the demand will emerge but rather when this will eventually occur.
- the ultimate market potential is still unclear, the estimates ranging widely.

These uncertainties are common for most technologies in their early development stages.

Rather than attempting to provide yet another set of estimates, it is probably more useful to address the issue of how to successfully manage these uncertainties, and take advantage of the opportunities offered by chirality.

At this aim, it is important to learn from past experiences drawn from similar areas. In this respect several interesting parallels can be made between the recent interest in chirality and the developments in genetic engineering and "modern biotechnology" that occurred in the mid 70's and early 80's.

Ten years ago, "modern biotechnology" had also attracted major interest and excitement from the academic community : almost all universities and research institutes have developed within a very short period departments active in this area, producing an overwhelming amount of literature on this subject. The industry has also been very quick in jumping into the band–wagon.

As a result, several billions of US$ have been invested in the area of genetic engineering and modern biotechnology innumerable start–ups being created to exploit the "unique" know–how of their founders.

Ten years later, it is interesting to have a retrospective view of what has really occurred in the area of modern biotechnology (see also figure 1) :

This set of technologies has proven to be an invaluable tool in a multitude of applications and industries. It has delivered, if not exceeded, the initial "realistic" expectations. Just a few examples :

- In the pharmaceutical industry genetic engineering is now a standard tool in the development of drugs, by allowing a more efficient screening of possible lead compounds through the development of drug–receptor binding assays made possible by the cloning of the appropriate receptors – a task that would have been very difficult using traditional techniques.

Similarly, new drugs have been made available through biotechnology such as for

example t–PA (tissue plasminogen activator) or EPO (erythropoietin);
- In agriculture the application of genetic engineering is paving the way for the development of new plant varieties with improved traits such as higher pest resistance and stress tolerance;
- In other more "pedestrian" fields the application of genetic engineering has improved the classical microbial strain optimisation process, increasing by several order of magnitude the productivity of some fermentation processes.

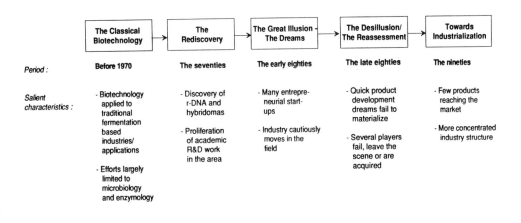

Figure 1. The development of biotechnology can be divided in five major stages.

The biotech scene has undergone a deep restructuring period in the late 80's, when the pendulum of expectations of quick returns from biotech eventually moved back into a more balanced situation with a more realistic assessment of the ultimate potential.

Not all the original actors have benefited equally from these development and a major shake–up has occurred in the biotech arena after the realisation of the hurdles involved.

While the success stories of biotech and genetic engineering flagships such as Genentech and Amgene are widely publicised, it is essential to remember that these represent the exception rather than the rule. Indeed, the number of failures in the field greatly exceeds the number of success stories !

Eventually, the companies who have succeeded are those that have been able to :

- integrate their technological capabilities within a broader set of skills, as clearly capabilities in genetic engineering and biotechnology alone are not enough in developing and launching a successful product,
- concentrate on selected niches or team–up with appropriate partners.

Similarly, most "end users", namely companies whose activities are potentially impacted by biotechnology, have developed in–house capabilities in this field, greatly reducing the opportunities open for independent players planning to sell their know–how.

These observations are raising similar questions in the area of chirality :

- Who will eventually take advantage of the potential opportunities available in this field?
- What strategies for success will need to be followed ?

AN OVERVIEW OF THE CHIRAL MARKET

Today a Small but Promising Market

The vast majority of synthetic pharmaceuticals, agrochemicals and aroma chemicals currently on the market and involving in their structure one or more centres of asymetry, are sold as mixtures of enantiomers or diastereoisomers (figure 2).

Only a handful of products are offered as pure enantiomers. The contrast with natural compounds where optical purity is the almost general rule (figure 3) is striking.

Molecular
Pool
(number of molecules)

Agrochemicals		Pharmaceuticals
± 700	total	> 2000
± 150	optically active*(a)	> 500
± 20	optically pure*(b)	< 100

Source : Arthur D. Little analysis

---

*	Excluding fermentation, extraction products and products obtained through semi-synthesis
(a)	With one or more centers of asymmetry.
(b)	Sold under the form of a pure enantiomer.

Figure 2. Relevance of optical purity in pharma and agrochem applications.
* Excluding fermentation, extraction products and products obtained through semi–synthesis.
(a) With one or more centers of asymmetry. (b) Sold under the form of a pure enantiomer.
Source : Arthur D. Little.

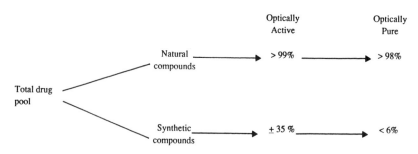

Figure 3. Nature indicates that optical purity must be important.

The current market for optically synthetic fine chemicals (intermediates and bulk active molecules) is therefore small and is less than US$ 300 Mio for intermediates and US$ 1000 Mio for bulk active molecules (see figure 4).

A)	Intermediates		B)	Bulk active compounds	
-	Amino acids[a]	60		Agrochemicals	200
-	Antibiotics side chains	120		mainly :	
-	Alpha-hydroxy acids [a]	20		- some pyrethroids	
-	Other	100		- flamprop isopropyl	
				Pharmaceuticals	800
				- captropril	
				- methyldopa	
				- naproxene	
				- timolol	
				- diltiazem	
				- enalapril	
				- dextrometophan	
	Total	+/- 300 Mio US$			+/- 1,000 Mio US$
		+/- 1,300 Mio US$			

(a) Intermediates applications only.
    Source : Arthur D. Little analysis

Figure 4. Current demand for chiral compound (1991). (a) Intermediates applications only. Source : Arthur D. Little.

The market is relatively concentrated, a few structures accounting for the bulk of the demand.

Although this market is small, it has attracted considerable interest as several driving forces are emerging, suggesting a major potential for expansion. These driving forces include :

• greater industry awareness
• evolving regulatory framework
• technological advances
• changing demand patterns

Greater Industry Awareness. The industry has only recently started to actively investigate the importance of optical purity.

Despite the overwhelming evidence shown by nature (figure 5) that only one of the enantiomers/diastereoisomers is biologically active, the industry has been satisfied to market products contaminated with a high proportion of isomeric ballast whose contribution ranges from useless at best, to noxious at worst.

In the pharmaceutical field it was only the Thalidomide tragedy that prompted the systematic investigation of the biological activity profile of the various optical isomers.

However, barring a few notable exceptions such as Methyldopa, Naproxen, Timolol, Diltiazem ... most drugs with a chiral center are still sold in the racemic form.

The "discovery" of optical purity is even more recent in the agrochemical field, and optical purity remains the exception, excluding in the pyrethoid segment.

In the past few years some agrochemical companies, ICI with its Fusilade herbicide being a prominent example, have been increasingly interested in developing optically pure pesticides.

Ecological concerns and the need to debottleneck production capacity have been the primary motivations behind this interest.

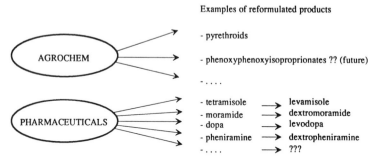

Figure 5. Reformulation of products currently sold under the racemic form will not be the general rule –mainly because of registration hurdles– similar to those imposed on an entirely new molecule.

Evolving Regulatory Framework. The industry is currently free to market a product in the optically pure form or as a racemic mixture – provided that no detrimental effects are demonstrated.

However, periodically it is rumoured that regulatory authorities will adopt a more rigid stance towards the marketing of pharma or agrochemical molecules as racemic of diastereoisomeric mixtures. Consensus seems to exist that the authorities will eventually call for the mandatory marketing of only the biologically active isomer, free of unwanted isomeric ballast.

Technological Advances. Until a few years ago the production of optically pure molecules was plagued by very high costs as the synthesis routes involved were either very cumbersome or plagued by low yields.

Recent advances have significantly improved overall economies. These include :

- the development of biocatalysts as suitable production tools either to resolve racemic mixtures (examples being the acylase system used by Degussa and Tanabe for the production of aminoacids or the hydantoïnase – carbamylase system applied by Recordati for the resolution of D/L para–hydroxy–phenylglycine) or as stereoselective synthesis tools (examples being the now defunct PAL–Phenylalamine Ammonia Lyase – of Genex or of the ICI benzene dihydroxylase systems)
- the economic application of *stereoselective chemical catalysts* like the hydrogenation systems developed by Monsanto and EniChem Sintesi for the synthesis of levodopa and L–phenylalanine
- the *availability of cheap chiral synthons* obtained often either by fermentation or as by-products of major processes and used as building blocks for further elaboration. Examples of such chiral synthons include lactic acid and tartrates.

Evolving Demand Patterns. The pharmaceutical and agrochemical industries are increasingly relying on a rational approach for the design of new molecules. This involves the careful analysis of how to optimise the interaction of the active molecule with the biological target in order to design the most in *vivo* effective molecule. This implies very often resorting to a chiral structure (this is not surprising as the living world is clearly asymmetric itself).

The ACE (angiotensin converting enzyme) inhibitors are good examples of a group of molecules issued from this type of approach.

Beware of Over–Optimism

In our view hopes, reported by some industry observers to reach rapidly a multibillion US$ market for chiral fine chemicals, are far too optimistic. Expectations need to be seriously reevaluated. The market will undoubtedly expand, but the timing and the extent of the expansion remain highly unclear and uncertain.

In assessing the future potential of this market, most estimates have failed to consider the following key aspects :

- the trend towards optical purity will be very gradual
- optical purity is unlikely to represent the general rule.

Optical Purity will not be the General Rule. Optical purity is often an unwanted luxury. The "inactive" isomer (frequently referred to as the distomer – the "bad" isomer) may have desirable properties (e.g. antagonistic properties, avoiding overdose complications, or scavenger effects protecting the active compound from rapid degradation like the sunlight). Also, often the inactive isomer may be racemised in vivo and converted in the active enantiomer (as for example ibuprofen).

It is highly improbable that products currently marketed in racemic form will be reformulated on a large scale as pure enantiomers. In the past this has been the case for only a handful of compounds (see figure 5) particularly as the registration hurdles and costs associated with the re–registration of the reformulated products are comparable to the development of a novel molecule. This can be illustrated by two recent examples :

- The reformulation of Fusilade under the optical pure form required the elaboration of a complete registration file
- Under the EINECS regulations (European Index of Existing Chemical Substances) the shipment or the sales of an optically pure enantiomer requires full toxicology and ecological characterisation of the enantiomer, even if the racemic mixture had been previously EINECS listed.

The Trend towards Optical Purity will be very Gradual. Contrary to common belief the percentage of optically pure synthetic pharmaceuticals and agrochemicals introduced in the market over the period 76–87 was remarkably constant (+/– 10% of the NCE (New Chemical Entities) introduced on the pharma market).

Looking ahead to the horizon of 1995–2000 we would expect an increase, possibly to 20–30% but certainly not as high as 50% – even if the registration procedure was dramatically tightened. Indeed :

- Active substances without any chiral centres will still be developed – witness the imnpressive series of new molecules developed by Janssen, one of the most productive designers of NCE, almost none of these having asymmetric centres.
- NCE in pharmaceuticals and agrochemicals take more than 7 years to develop. Therefore, even if the industry decided today to develop only optically pure

compounds, this decision would have an impact on the market well until after 2000.

Overall a Limited Market Potential

Considering all these elements and knowing that the new molecules – introduced over the past ten years – account today for less than 25% of all industry sales – the rapid creation of larger demand for optically pure fine chemicals is to be ruled out.

Furthermore hopes of a larger demand of chiral fine chemicals in new applications such as liquid crystals are also unrealistic, particularly in light of the very limited volumes involved (total world consumption of nematic liquid crystals – by far the most commercially interesting – being less than 20 t/y).

Therefore in our view, the market for chiral molecules, today probably in the range of US$ 1,200 Mio, is unlikely to exceed significantly US$ 2.5 Bio even at the Horizon 2000 (see figure 6).

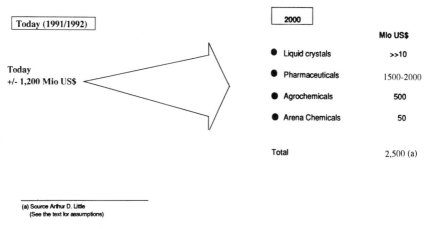

(a) Source Arthur D. Little
(See the text for assumptions)

Figure 6. What ultimate potential for optically pure fine chemicals ? (see the text for assumptions)
(a) Source : Arthur D. Little.

THE STRATEGIC MANAGEMENT OF CHIRALITY

In order to analyse how to best harness the potential offered by chirality it is important to analyse the current stage of technological development in this area.

Five main families of technologies can be identified in the field of chirality :

- classical chemical resolution
- enzymatic resolution
- stereoselective chemical synthesis
- stereoselective biochemical synthesis
- use of optically pure synthons (obtained using one of the above mentioned techniques).

These techniques are in different stages of maturity in terms of :

- range of industrial applications

- expected potential
- ....

as illustrated in figure 7 (table 2 providing a brief reminder of the elements used in assessing the degree of maturity of these various techniques).

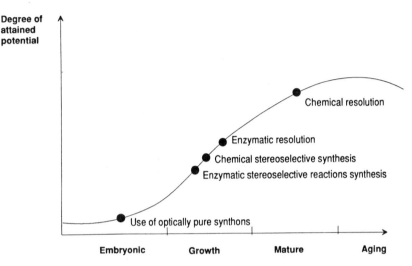

Figure 7. Maturity of various chiral technologies.

Similarly, the level of competitive impact of these various techniques varies widely, depending on the industrial application considered (see table 3 for the definitions used in assessing the competitive impact).

Table 3. Competitive impact of technologies (Source : Arthur D. Little)

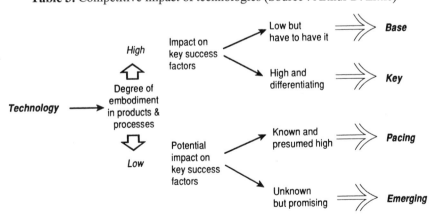

As an example, while chemical resolution is a very well established technique that can be considered as "base" at least in the pharmaceutical industry, the impact of stereoselective chemical synthesis in the area of high performance polymers is still highly speculative.

Table 2. Technology maturity determination (Source : Arthur D. Little)

Stages of technology maturity / Descriptor	Embryonic (E)	Growth (G)	Mature (M)	Aging (A)
Technological Development	Discovery	Adaptation	Augmentation	Stagnation
Competitive situation	Indeterminate	Shake-out	Demand led	Stabel/defensive
Ease of entry	Specialization	Cooperative	Licensing	Wholesale
Emergence of standards	Advance promotion	Producer conflicts	Producer volume	Bureaucratic
Product development	Experimentation	Batch production	Mass production	Cosmetic
Application areas	Research	Leading edge	Mass market	Basic necessity
Growth & penetration	Not applicable	Adoptive	Explosive	Stabilization
Slope of unit cost curve	Indeterminate	Falling steeply	Flattening	Flat
Costs	Basic R&D = X	Product development = 10X	Market development = 100X	Legal defence/technology replacement=?
Price premium/advantage	Not known	High-medium	Medium-low	Possible discount
Publisher channels	Scientific journals	Trade press	Daily papers	

Source : Arthur D. Little

This analysis, combined with the expected impact of chirality in various end user markets (see figure 8) is summarised in table 4.

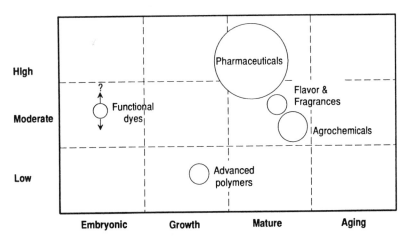

Figure 8. Likely impact of chirality on various end user industries.
N.B.: Circle size roughly proportional to industry turnover. Source : Arthur D. Little.

What are the implications of this analysis and what do these mean in practice ?

Some technological capabilities are now "base" and offer limited scope for competitive advantage. As an example by and large the field of chemical resolution, is now a well established tool in several applications and industries and does not offer major scope for competitive edge. It is therefore important avoiding overinvesting in it.

On the other hand, stereoselective enzymatic synthesis has still an unclear competitive impact and is a field worth monitoring actively. Should breakthroughs be achieved in this area the pay–offs should be considerable, offering rich pickings for innovative companies.

This clearly indicates that there is a need for a careful and highly differentiated approach in the technology investment decision.

This analysis has also important implications in terms of opportunities for sourcing of the technology. Indeed, it is very common for emerging and base technologies to be outsourced (although the outsourcing decision is based on completely different considerations : for emerging technologies not to immobilise scarce resources in an area of yet unclear potential while for base technologies the outsourcing decision is often based on overall efficiency considerations). This is much less the case for pacing and key technologies, given their possible role in building competitive edge.

Similarly, it is important to clearly understand where the value added lies in the product development process and what capabilities need to be mastered in order to succeed.

Indeed, several start–up companies in the field of modern biotechnology have originally envisaged emerging as full fledged pharmaceutical companies but have eventually failed as they were mastering only one element in the product development cycle.

Within this line of thought, it is essential to realise that chirality represents only a set of enabling technologies.

Table 4. The competitive impact of the various chiral technologies differs depending on the application considered (Source : Arthur D. Little).

Technology / Application	Optically pure synthons	Enzymatic stereoselective synthesis	Chemical stereoselective synthesis	Enzymatic resolution	Chemical resolution
Pharmaceuticals	▦ Pacing	▧ Key	▧ Key	▧ Key	▓ Base
Flavor and fragrances		▦ Pacing	▦ Pacing	▦ Pacing	
Agrochemicals	▧ Key				▧ Key
Advanced polymers			≡ Emerging		≡ Emerging
Functional dyes			≡ Emerging		≡ Emerging

≡ Emerging   ▦ Pacing   ▧ Key   ▓ Base

Source : Arthur D. Little

   Although interesting and of high relevance to several industries, these technologies represent only one of the skills that need to be mastered. As an example in the pharmaceutical industry mastering the chiral step would be of limited use per–se as it requires much more to get the final molecule, not to mention the formulated product!

   Similarly, it is important to realise where the value added lies. It is very easy to fall in the trap consisting of thinking that the market for chiral fine chemicals will reach several billion US$ based on the argument that at least 20–30% of all new drugs will be optically pure at the horizon 2000 and that just a 1% of the drug market corresponds to one or two billion US$! This is completely overlooking that :

- The end product price is a multiple of the cost of the active ingredient (a multiplier of 10 to 20 being far from uncommon)
- The value of the chiral step represents only a fraction of the cost of the active ingredient!

   These considerations are raising once again the issue of integration and what an independent supplier offering "capabilities in chiral synthesis" can hope to achieve.

CONCLUSIONS

   The field of chirality is attracting a considerable interest and seems to be promised significant developments.

   However, successful participation to this field calls for several comments and requires addressing key issues including :

- level of spending and efforts
- level of back and upstream integration
- position in the activity chain targeted.

The ultimate answer to these questions will of course vary depending on the type of industry participant.

Integrated industry players in areas where chirality is expected to play a significant role, such as pharmaceuticals, will undoubtedly need to continue developing capabilities in this field, at least at the R&D level.

Corporate or academic liaisons could be envisaged for securing an access to emerging techniques, that may have a significant competitive impact in the future but have still high levels of uncertainty associated with.

Similar considerations apply to the industry players in areas such as polymers and functional dyes where the impact of chirality is still unclear. Typically, they would be better off by outsourcing their chiral technologies.

Base technologies will be typically mastered at the full production scale in–house, although there could be scope for outsourcing these to partners having developed better levels of overall efficiency.

Established fine chemicals suppliers will, similarly, need to develop capabilities in the chiral field, where we expect the demand for such services to increase at the industrial scale and to become a standard tool in custom or even toll production.

Probably, the toughest challenges in successfully harnessing the opportunities offered by chiral synthesis will be faced by the technology start–ups. While these certainly represent the best source for breakthroughs and innovation we do not believe that most start–ups will be able to survive except if they eventually join forces with appropriate corporate partners.

STRATEGIES FOR ASYMMETRIC SYNTHESIS. WHAT IS THE ROLE OF HETEROGENEOUS CATALYSIS ?

Léon Ghosez

Université Catholique de Louvain
Laboratoire de Chimie Organique de synthèse
place Louis Pasteur, 1
B – 1348 Louvain–la–Neuve, Belgium

ABSTRACT

Enantioselective synthesis is and will remain without any doubt one of the most active area in synthesis. The reason is that many properties of molecules which have potential use as pharmaceutical or agrochemical products or as new materials depend on their absolute configuration. It is thus necessary to develop efficient and economical methods which can be applied to the synthesis of complex molecular systems and meet the following stringent criteria : (a) high efficiency, (b) low cost, (c) friendliness to the environment.

Reactions involving at least one enantiomerically pure reagent or catalyst can lead to an enantiomerically pure product if they occur with total facial selectivity. It is practical to provide the chiral information through a molecular fragment temporarily bonded to one of the reagent or, ideally, to a catalyst (scheme 1).

At present most asymmetric syntheses rely upon the use of a covalently bonded chiral auxiliary. This requires at first the attachment of the auxiliary to one of the reagent by a covalent bond which can be readily cleaved at the end of the sequence. The cleavage step should take place in high yields and without epimerization of the various chiral centers. The chiral auxiliary should be recovered in high yields and without racemisation. Highly successful and practical asymmetric syntheses following this strategy have been described in recent years. Some typical examples are shown in scheme 2.

In recent years one has seen several interesting examples of reactions using Lewis acids bearing enantiomerically pure ligands. These act as catalysts by coordination to a polar substituent (often a carbonyl group) of the electrophilic partner of the reaction and, at the same time, they provide the chiral information. In many cases they must be used in large if not stoechiometric amounts. Thus they cannot be regarded as true catalysts but rather as chiral activators. An illustration of this strategy is shown in scheme 3 [6].

Asymmetric catalysis represents the ultimate solution for the preparation of enantiomerically pure compounds. Only a small amount of the often expensive chiral would

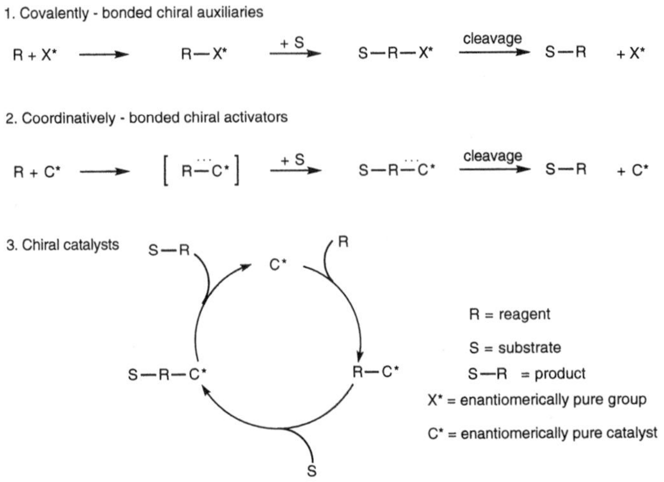
1. Covalently - bonded chiral auxiliaries
R + X* → R—X* + S → S—R—X* cleavage → S—R + X*
2. Coordinatively - bonded chiral activators
R + C* → [R···C*] + S → S—R···C* cleavage → S—R + C*
3. Chiral catalysts
R = reagent
S = substrate
S—R = product
X* = enantiomerically pure group
C* = enantiomerically pure catalyst

Scheme 1.

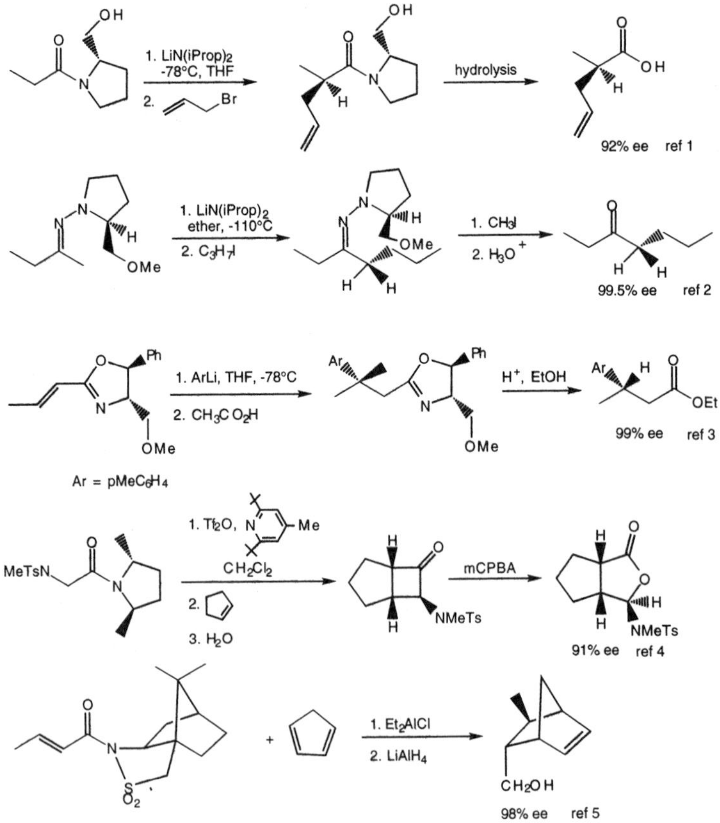
1. LiN(iProp)2 -78°C, THF
2. Br
hydrolysis
92% ee ref 1
1. LiN(iProp)2 ether, -110°C
2. C3H7I
1. CH3I
2. H3O+
99.5% ee ref 2
1. ArLi, THF, -78°C
2. CH3CO2H
H+, EtOH
99% ee ref 3
Ar = pMeC6H4
1. Tf2O,
CH2Cl2
2.
3. H2O
mCPBA
91% ee ref 4
1. Et2AlCl
2. LiAlH4
98% ee ref 5

Scheme 2.

be needed and the desired product would be directly formed. The requirements for a good asymmetric catalyst are very stringent : (a) it should lead to a very high facial selectivity because the purification of mixtures of enantiomers can sometimes be difficult, (b) its turn-over should also be high. These requirements are very difficult to meet as a result of the difficulty to rationally devise a chiral catalyst and of the often insufficient knowledge of the catalytic cycle. Nevertheless, in recent years, spectacular examples of truly catalytic asymmetric reactions have been described. Sharpless' epoxidation of allylic alcohols undoubtedly represents a breakthrough in the field (scheme 4) [7].

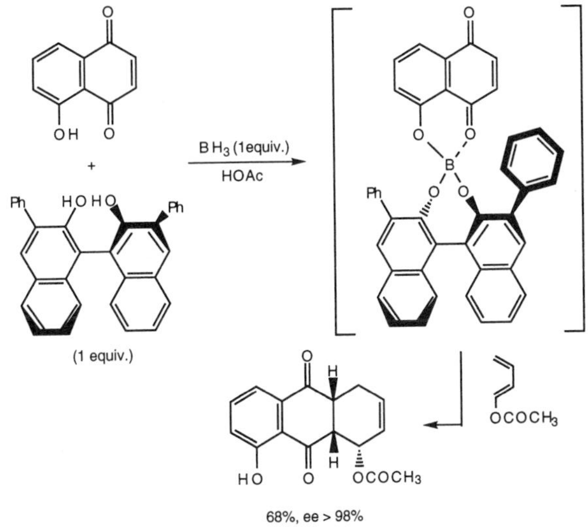

68%, ee > 98%

Scheme 3.

70-90% Yield
>90% ee

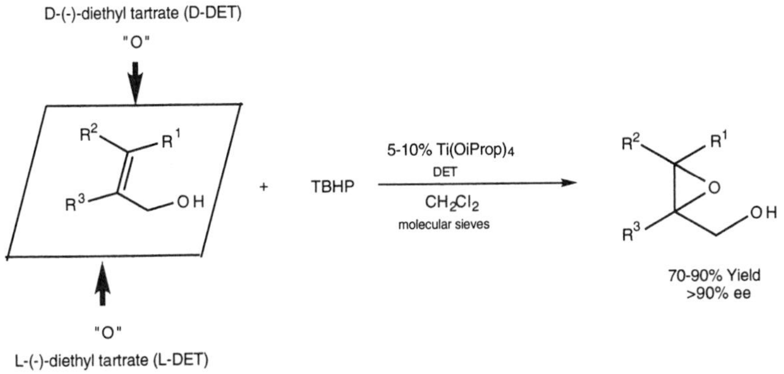

Scheme 4.

The reaction is truly catalytic, uses cheap reagents and produces useful synthetic intermediates in both enantiomerically pure forms. However it is limited in scope. A more general approach to the catalytic asymmetric oxidation of olefins has been described by the same group (scheme 5) [8].

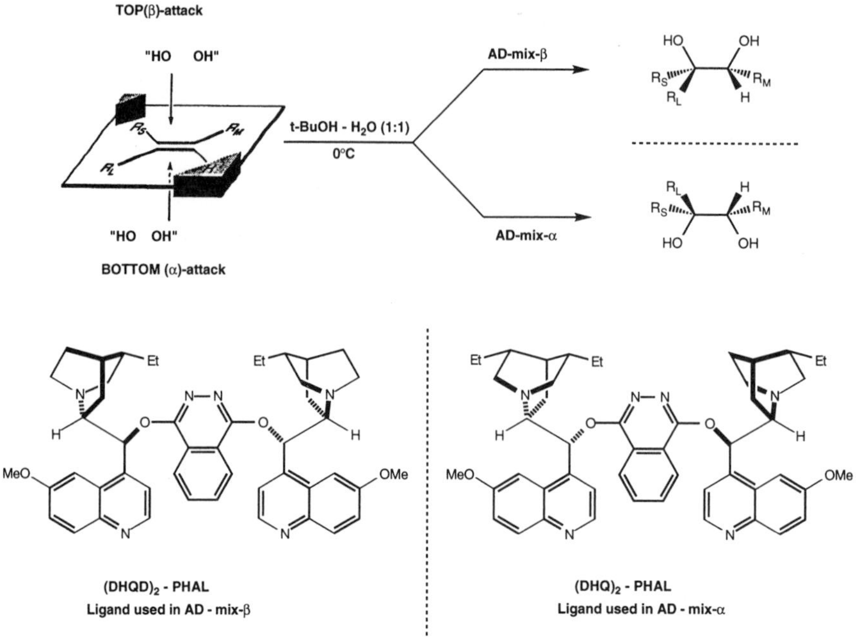

Scheme 5.

Asymmetric catalytic reactions which lead to the formation of carbon–carbon bonds are key processes in organic synthesis. The cycloaddition of ketene to chloral catalysed by quinidine is an elegant illustration of the standards which should be attained (scheme 6) [9].

However similar selectivities in the synthetically more important cycloadditions of ketenes to olefins and imines have not been obtained.

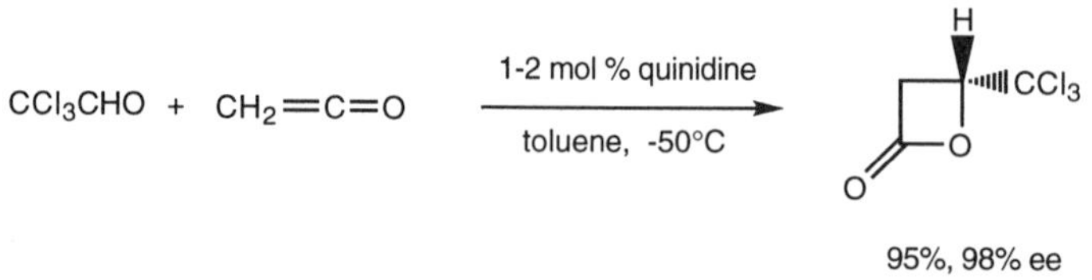

95%, 98% ee

Scheme 6.

Another significant development of asymmetric catalysis for the formation of C–C bond is shown in scheme 7. The coordination of organozinc compounds to chiral aminoalcohols result in an activation of the organometallic species which readily add to aldehydes. This process requires only catalytic amount of the optically pure aminoalcohol. The turn–over is very high [10].

In the next few years we shall certainly see significant progresses in the ability of organic chemists to design new catalysts using an "enzymomimetic" approach. The use of secondary attractive catalytic–substrate interactions to improve facial selectivity has recently been illustrated (scheme 8) [11].

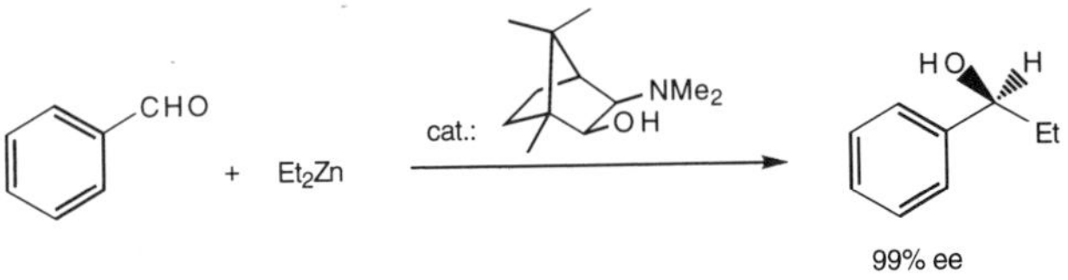

Scheme 7.

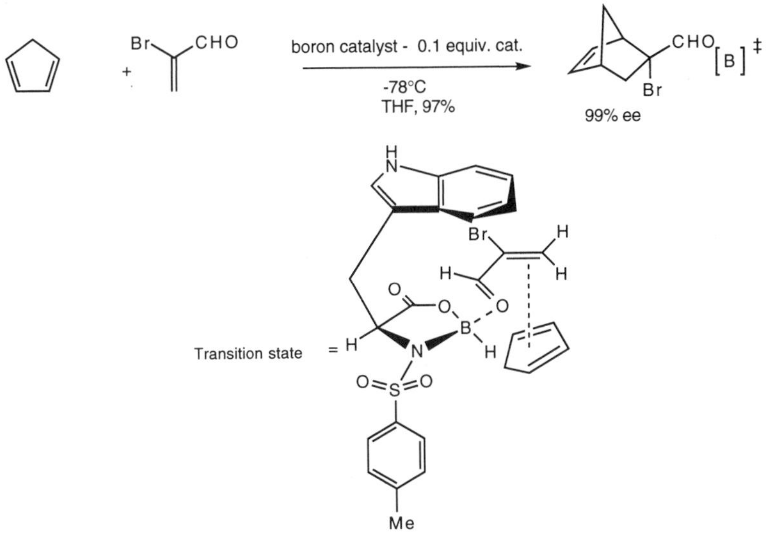

Scheme 8.

For many years, synthetic chemists have been using biochemical catalysts in both industrial and laboratory processes. Microorganisms, yeast, enzymes are now part of the armamentarium of synthetic chemists.

It should however be remembered that nature usually provides catalysts for the production of only one of the enantiomers. Furthermore biocatalysts are more substrate sensitive than their chemical counterparts. Finally it should be remembered that nature has generated its catalysts to support life. Obviously there are many reactions which do not exist in nature and, therefore, for which there is no biocatalyst. There is some hope that, with the advances of genetic engineering, unnatural enzymes with a desired catalytic activity could be generated, isolated and produced in significant amounts. One can also use the immune system to produce biocatalysts. Thus a "Diels–Alderase" has been generated by submitting the immune system to an hapten mimicking the transition state of the reaction (scheme 9) [12].

Other illustrations of the successes of homogeneous enantioselective catalysis sing transition metal complexes will be described in the lecture of H. Brunner.

What can be expected from heterogeneous catalysis for the production of enantiomerically pure compounds ? The development of such processes would be extremely useful because heterogeneous catalysts can be easy separated from reaction mixtures and can be reactivated for further use. This would solve an environmental problem resulting from the fact that recycling an homogeneous catalyst with respect to the metal and even more to the chiral ligand can be problematic and anyway very costly.

It is more difficult to synthesize chiral heterogeneous catalysts than the corresponding homogeneous catalysts [13]. Also the design of an heterogeneous catalyst providing a desired selectivity is often impossible because of a lack of understanding the mechanism with sufficient precision. Nevertheless some efficient examples of heterogeneous enantioselective catalysis are known. I will focus here on the few examples in which C–C bonds are formed enantioselectively. Up to now it has been possible to design a chirally modified heterogeneous catalyst to create new C–C bonds. Successes have been obtained, however, for hydrogenation and oxidation reactions.

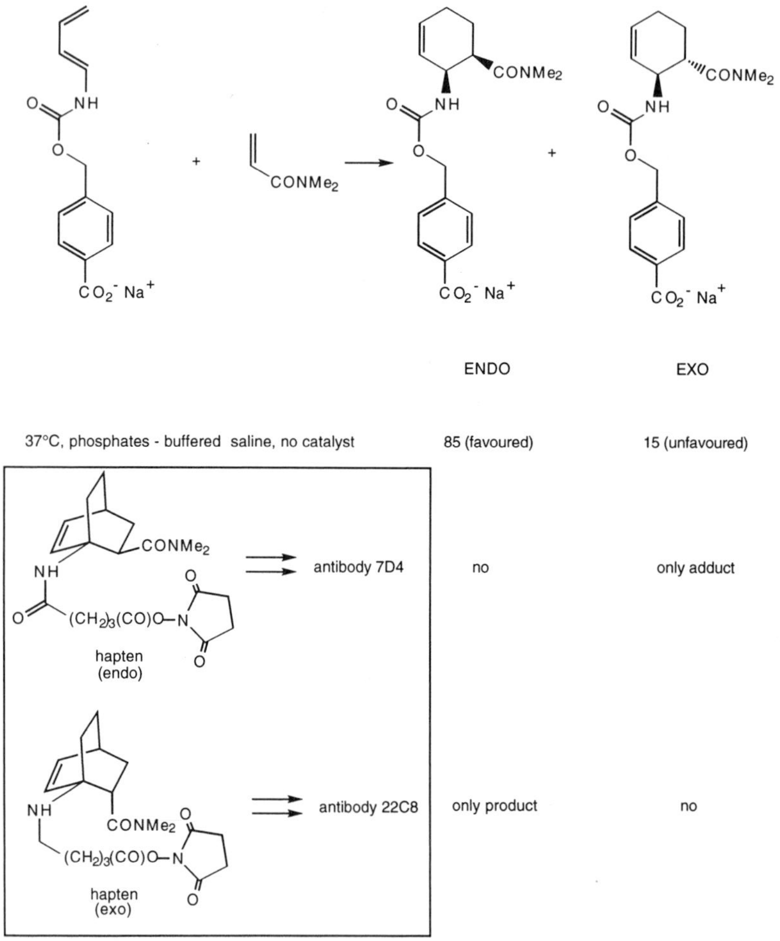

Scheme 9.

Heterogenisation of homogeneous catalysts appear to be more promising and a number of asymmetric syntheses using polymer–bound metal complex catalysts have been reported. There are still many problems to be solved : many immobilized molecular catalysts are less reactive than in solution because of decreased molecular motions. Easy recovery of the catalyst is not always as easy as it appears : most polymer–bound organometallic catalysts deteriorate after a few uses. The examples shown in scheme 10 are among the best in the field. Most often enantioselectivities are mediocre and there are problems with the activity of the catalyst. I do not foresee that heterogeneous catalysis will become a much used technique for the formation of C–C bond at least in the near future. More research is still needed which would allow a more efficient design of these catalysts.

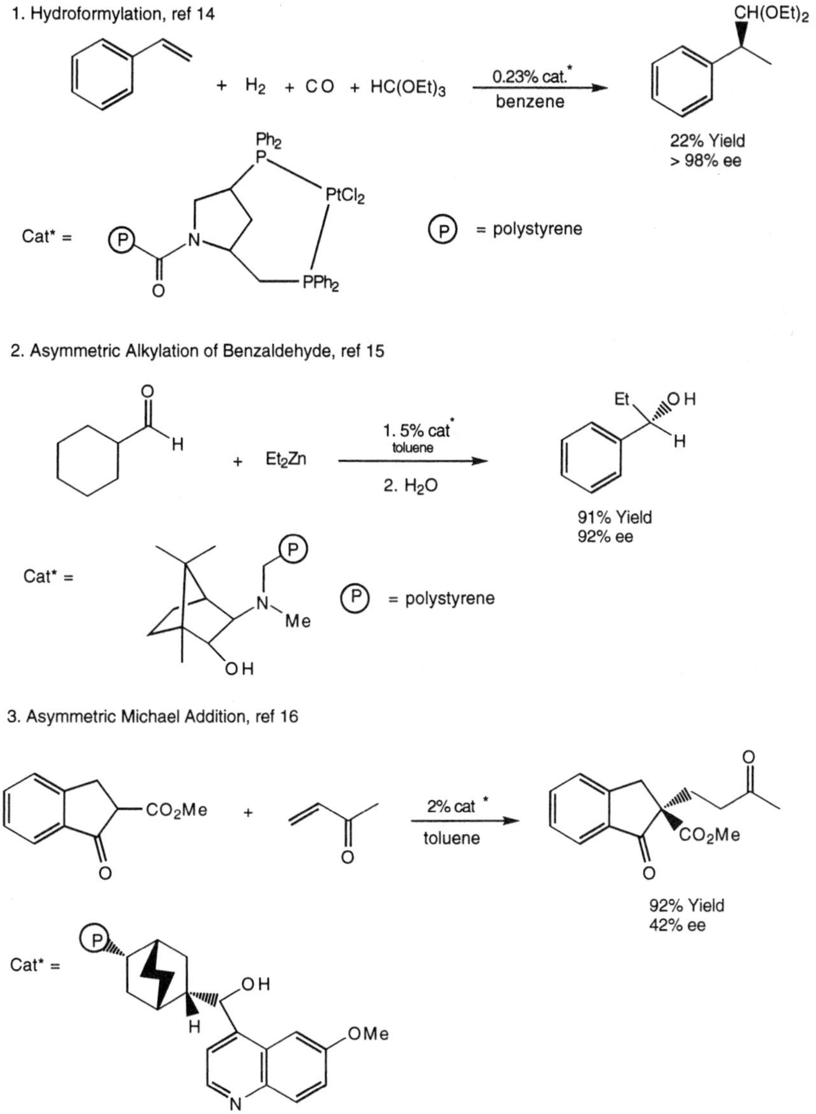

1. Hydroformylation, ref 14

2. Asymmetric Alkylation of Benzaldehyde, ref 15

3. Asymmetric Michael Addition, ref 16

Scheme 10.

REFERENCES

[1]   Evans, D.A.; Takacs, J.M. *Tetrahedron Lett.* **1980**, *21*, 4233.

[2]   Enders, D.; Eichenauer, H.; Baus, U.; Schubert, H.; Krener, K.A.M. *Tetrahedron*, **1984**, *40*, 1345.

[3]   Meyers, A.I.; Smith, R.K. *Tetrahedron Lett.* **1979**, *20*, 2749.

[4]   Genicot, C.; Ghosez, L. *Tetrahedron Lett.* **1992**, *33*, 7357.

[5]   Oppolzer, W. *In Comprehensive Organic Synthesis*, Trost, B.M.; Fleming, I. ed., Pergamon Press, **1991**, *5*, chap. 4.1.

[6]   Kelly, T.R., Whiting, A.; Chandrakumar, N.S. *J. Amer. Chem. Soc.* **1986**, *108*, 3510.

[7]   Reviews : (a) *Synthetic Aspects and Applications of Asymmetric Epoxidation,* vol. 5, pp. 194–246, Rossiter B.E. in J.D. Morrison, ed., Asymmetric Synthesis, Academic Press, Orlando, Fl. (1985); (b) *Asymmetric Epoxidation of Allylic Alcohols : The Sharpless Epoxidation,* Pfenninger A., Synthesis, 89–116 (1986); (c) Johnson R.A. and Sharpless K.B. in *Comprehensive Organic Synthesis,* vol. 7, chap. 3.2, Trost, B.M.; Fleming I. and Ley, S.V., eds, Pergamon Press, Oxford, U.K. (1991).

[8]   Sharpless, K.B.; Amberg, W.; Bennani, Y.L.; Crispino, G.A.; Hartung, J.; Jeong, K.–S.; Kwong, H.–L.; Morikawa, K.; Wang, Z.–M.; Xu, D. and Zhang, X.–L. *J. Org. Chem.* **1992**, *57*, 2768–2771.

[9]   Wynberg, H.; Staring, E.G.T. *J. Amer. Chem. Soc.* **1982**, *104*, 166; Wynberg, H.; Staring, E.G.T. *J. Org. Chem.* **1985**, *50*, 1977; Ketelaar, P.E.F.; Staring, E.G.T.; Wynberg, H. *Tetrahedron Lett.* **1985**, *26*, 4665.

[10] Noyori, R.; Kitamura, M. *Angew. Chem. Int. Ed.* **1991**, *30*, 49.

[11] Corey, E.J.; Loh, T.P. *J. Amer. Chem. Soc.* **1991**, *113*, 7794.

[12] Gouverneur, V.E.; Houk, K.N.; de Pascual–Teresa, B.; Beno, B.; Janda, K.D.; Lerner, R.A. *Science* **1993**, *262*, 204.

[13] Noyori, R. in *Asymmetric Catalysis in Organic Synthesis,* chap. 8, 346, J. Wiley and Sons, Inc. New–York, 1994.

[14] Stella, J.K. *J. Macromol. Sci., Chem* **1984**, *A 21*, 1689; Parinello, G.; Stella, J.K. *J. Amer. Chem. Soc.* **1987**, *109*, 7122.

[15] Itsano, S.; Frechet, J.M.J. *J. Org. Chem.* **1987**, *52*, 4140.

[16] Kobayashi, N.; Iwai, K. *J. Amer. Chem. Soc.* **1978**, *100*, 7071; Kobayashi, N.; Iwai, K. *J. Polym. Sci.; Polym. Chem. Ed.* **1980**, *18*, 923.

HOMOGENEOUS ENANTIOSELECTIVE CATALYSIS

Henri Brunner

Universität Regensburg
Institut für Anorganische Chemie
D–93040 Regensburg, Germany

ABSTRACT

The intention of the lecture on homogeneous enantioselective catalysis in the ChiCat Conference on heterogeneous enantioselective catalysis was to work out the differences between homogeneous and heterogeneous enantioselective catalysis and to show the contributions of homogeneous catalysis to the understanding of a catalyst's enantio-selectivity.

The most celebrated reaction in enantioselective catalysis with transition metal compounds is the hydrogenation of prochiral olefins, specifically of dehydroamino acids to give amino acids [1]. Scheme 1 shows the hydrogenation of Z–α–acetamido cinnamic acid to N–acetylphenylalanine.

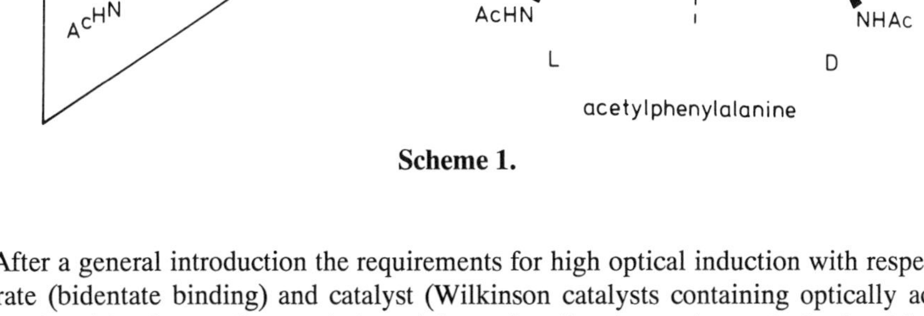

Scheme 1.

After a general introduction the requirements for high optical induction with respect to substrate (bidentate binding) and catalyst (Wilkinson catalysts containing optically active chelate phosphines) were discussed. A model to rationalize the results on the basis of the δ/λ

conformation of puckered five–membered chelate rings was given which represents an approach typical for homogeneous enantioselective catalysis [2]. It was pointed out that the equatorial/axial arrangement of the face–exposed/edge–exposed phenyl rings of the $P(C_6H_5)_2$ groups transmits the chiral information within the hydrogenation catalyst over a distance of ca. 5 Å. The famous 3 kcal of stereochemistry and its relevance to molecular modelling was discussed.

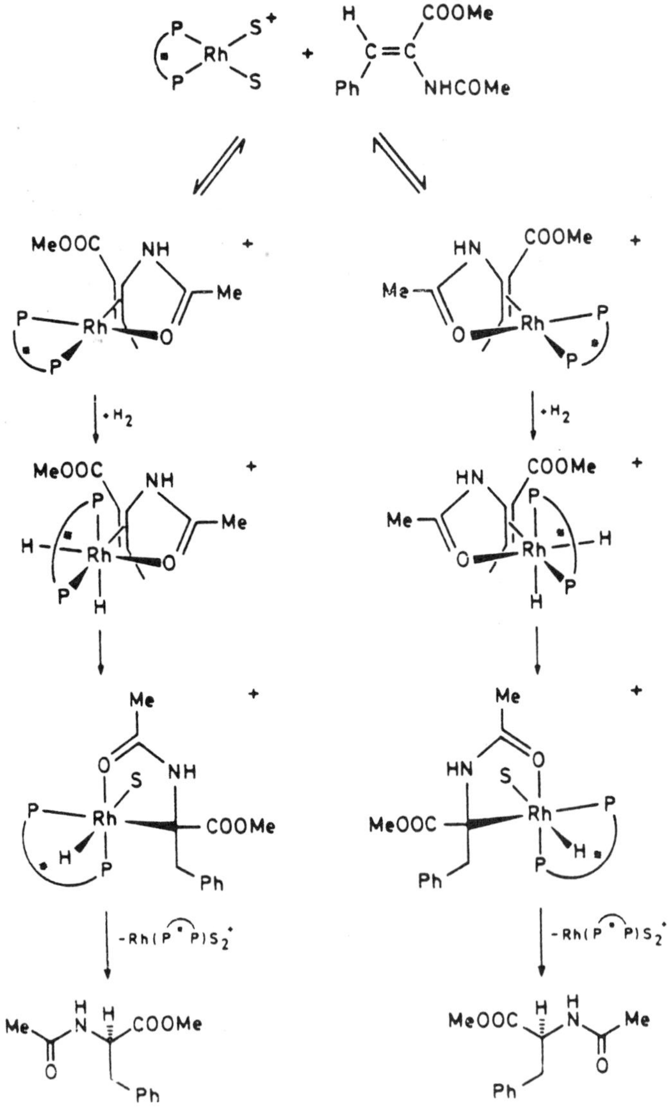

Scheme 2.

In homogeneous enantioselective catalysis, kinetic measurements, spectroscopic studies etc. can elucidate a reaction mechanism, at least in favorable cases such as the hydrogenation of dehydroamino acids. As an example, the famous Halpern/Brown mechanism was presented (scheme 2). There is an equilibrium between two diastereomeric complexes containing the optically active chelate phosphine and the bidentate dehydroamino acid derivative. This equilibrium is dominated by the major isomer, the minor isomer being present only in amounts of a few per cent. The rate determining step is the oxidative addition of hydrogen which is followed by the insertion of the olefin into the Rh–H bond and by the reductive elimination of the product. Surprisingly, due to its high reactivity, the minor diastereomer of the diastereomer equilibrium is the product forming species [3].

In heterogeneous enantioselective catalysis it is impossible to establish mechanisms in such detail as shown for the homogeneous hydrogenation of dehydroamino acids. However, a warning was added not to generalize such mechanistic studies. It was mentioned that the hydrogenation of cyclohexene by the actual Wilkinson catalyst $RhCl(PPh_3)_3$ follows a completely different mechanism, including intermediates with PPh_3 ligands in trans position, oxidative addition of hydrogen preceding the activation of the substrate and a shift of the rate determining step to the insertion of the olefin into the Rh–H bond.

Another reaction type in homogeneous enantioselective catalysis is the hydrosilylation of prochiral ketones. Scheme 3 shows the reaction of acetophenone with diphenylsilane. Under the influence of a catalyst, the Si–H bond adds across the C=O bond to give a silylether. In a subsequent step, this silylether is hydrolyzed, the final product being 1–phenylethanol (scheme 3).

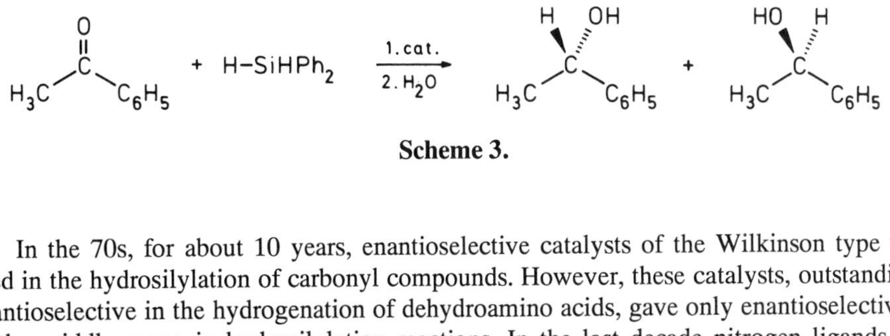

Scheme 3.

In the 70s, for about 10 years, enantioselective catalysts of the Wilkinson type were used in the hydrosilylation of carbonyl compounds. However, these catalysts, outstandingly enantioselective in the hydrogenation of dehydroamino acids, gave only enantioselectivities in the middle range in hydrosilylation reactions. In the last decade nitrogen ligands, e.g. pyridine imines, pyridine thiazolidines, and pyridine oxazolines, were developed which turned out to be more efficient ligands than optically active phosphines [4]. Thus, a kind of catalyst–specificity is arising. For each reaction type and for each reaction the best enantioselective catalyst has to be found.

In addition to the hydrogenation of dehydroamino acids and the hydrosilylation of prochiral ketones there are many different reaction types, catalyzed by transition metal compounds with high enantioselectivities. This information is accumulated in the author's new Handbook of Enantioselective Catalysis with Transition Metal Compounds [5]. However, there are also reaction types which even today are unsolved problems. One example is the Pd–catalyzed allylation of soft nucleophiles (scheme 4, right side). As the nucleophile attacks the π–allyl ligand in the intermediate from the side opposite to the metal atom (scheme 4, left side), conventional optically active phosphines are too small to induce enantioselectivity at the α–C atom of the nucleophile in the formation of the new C–C bond [6]. As a concept to solve this problem the synthesis of expanded ligands was suggested. Scheme 5 (left side) shows in a general way how to a chelating PP backbone different layers of chiral or achiral substituents can be added to expand the ligands. In scheme 5 (right side) a specific example of a two–layer ligand is given, carrying 8 menthyl groups at the outside [7]. With this approach, it should be ultimately possible to carry out enantioselective reactions such as the allylation shown in scheme 4 in catalyst–structures resembling the clefts and pockets of enzymes.

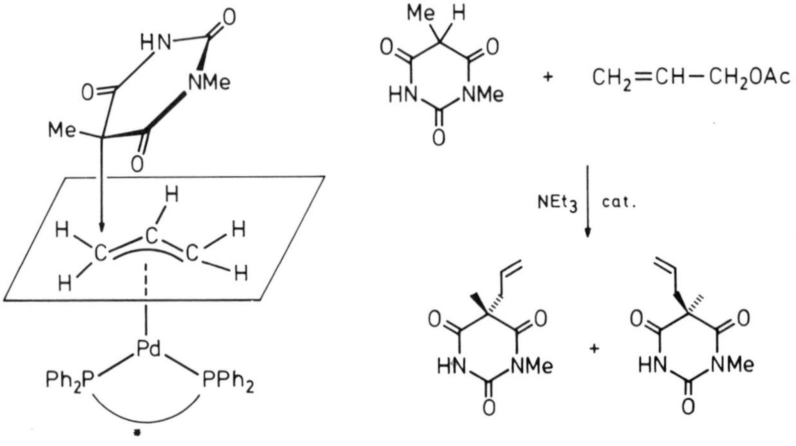

Scheme 4.

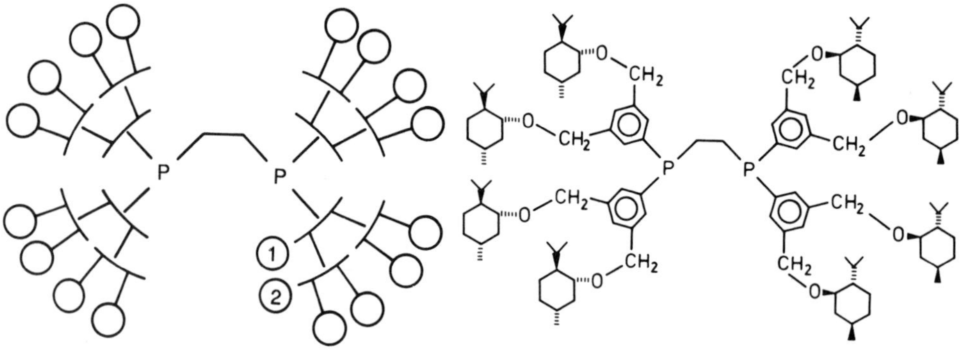

Scheme 5.

REFERENCES

[1] K. E. Koenig in J. D. Morrison: "Asymmetric Synthesis", Vol. 5, Academic Press, Orlando 1975, p. 71.

[2] W. S. Knowles, B. D. Vineyard, M. J. Sabacky, B. R. Stults in M. Tsutsui: "Fundam. Res. Homogeneous Catal.", Vol. 3, Plenum Press, New York 1979, p. 537.

[3] J. Halpern in J. D. Morrison: "Asymmetric Synthesis", Vol. 5, Academic Press, Orlando 1975, p. 41.

[4] T. Hayashi in I. Ojima: "Catalytic Asymmetric Synthesis", VCH, New York 1993, p. 325.

[5] H. Brunner, W. Zettlmeier, "Handbook of Enantioselective Catalysis with Transition Metal Compounds", VCH Verlagsgesellschaft, Weinheim 1993.

[6] H. Brunner, H. Nishiyama, K. Itoh in I. Ojima: "Catalytic Asymmetric Synthesis", VCH, New York 1993, p. 303.

[7] H. Brunner, J. Fürst, J. Ziegler, J. Organomet. Chem. 454 (1993) 87.

SCOPE AND LIMITATIONS OF THE APPLICATION OF HETEROGENEOUS ENANTIOSELECTIVE CATALYSTS

Hans-Ulrich Blaser and Benoît Pugin

Ciba-Geigy AG,
Central Research Services, R 1055.6
CH-4002 Basel, Switzerland.

SUMMARY

The emphasis of this overview is placed on the synthetic potential and relevant characteristics of three classes of synthetic heterogeneous chiral catalysts (not covered are heterogeneous biocatalysts). i) *Modified "classical" heterogeneous catalysts.* The addition of chiral, low molecular weight modifiers to supported metal catalysts is most successful for preparing hydrogenation catalysts for α- and β-functionalized ketones. The main emphasis is on synthetic and technical applications but an update on mechanistic investigations and new systems is also given. The use of modified metal oxides is summarized. So far, only modified pillared clays for the epoxidation of allylic alcohols and the oxidation of aryl sulfides are of synthetic interest. ii) *Chiral polymers.* Reviewed are the uses of natural and synthetic chiral polymers as catalysts and supports. A few interesting concepts are described even though the synthetic application is restricted to poly(aminoacids) as selective catalysts for the epoxidation of chalcones and a polymer supported dipeptide for the hydrocyanation of aldehydes. A short section describes enantioselective reactions in chiral liquid crystal and gels. iii) *Immobilized metal complexes.* The preparation of different types of immobilized analogs of effective homogeneous chiral catalysts is assessed in detail. Each type is illustrated with representative examples. Requirements are discussed that are thought to be decisive for their synthetic application. Immobilized catalysts with high enantioselectivity and satisfactory activity have been developed for several reaction types. Most work has been carried out in hydrogenation, hydroformylation and oxidation reactions.

INTRODUCTION

From an industrial point of view, heterogeneous as opposed to homogeneous catalysts have the inherent advantage of easy separation and very often also of better handling properties. Most enantioselective reactions described in the literature, however, are carried out homogeneously using stoichiometric or catalytic amounts of a chiral auxiliary [1,2]. In this overview we will discuss scope and limitations of different types of heterogeneous catalyst systems for enantioselective reactions. The emphasis will be on their potential for synthetic applications but new results on other aspects will be summarized as well. We will not cover the use of immobilized enzymes or organisms even though these catalysts are also heterogeneous and can be used for synthetic purposes [2,3].

Three general approaches for preparing chiral heterogeneous catalysts will be discussed: i) The chiral modification of "classical" heterogeneous catalysts; ii) the use of chiral polymers and iii) the immobilization of chiral homogeneous catalysts. Each of these strategies will be illustrated by the best examples reported in the literature. In addition, evaluation criteria are discussed and an attempt will be made to assess the different catalytic systems.

MODIFIED "CLASSICAL" HETEROGENEOUS CATALYSTS

With this strategy, a classical heterogeneous catalyst with the required activity for a desired transformation is modified with a small chiral organic molecule. This is a very flexible approach because many different types of structurally well defined modifiers are available. Also, the modifier can be added to the catalyst or to the reaction solution. Nevertheless, not many modified catalysts have been found with really high enantioselectivity. The major reason for this fact is probably the lack in understanding of the mode of action of such catalysts, i.e., one does not known how the enantioselection works. Therefore, new modified catalysts for new reaction types will rather be found empirically and not by design.

Chirally Modified Hydrogenation Catalysts

Several recent reviews give an excellent digest of the most relevant aspects of chirally modified catalysts [4,5]. Therefore, our description will be restricted to the description of some synthetic applications and a summary of results reported in the last two years.

Tartrate-Modified Nickel and Related Catalysts. The development and the successful application of the Nickel-tartrate system as enantioselective hydrogenation catalysts have been reviewed by Tai and Harada [6]. The preparation of the modified Nickel catalysts on a larger scale is somewhat delicate and requires careful control of the modification conditions. A tartrate modified Raney Ni catalysts is now also commercially available [4a].

Synthetic Applications. A multistep synthesis of several isomers of the sex pheromone of the pine sawfly starts with the nickel catalyzed hydrogenation of methyl 2-methyl-3-oxobutyrate with fair stereoselectivity [7]. Recently, a new synthesis was published starting with the enantioselective hydrogenation of methyl acetoacetate with the same catalytic system [8]. Biologically active C_{10}-C_{16}-3-hydroxyacids were produced with optical yields of 83-87% starting from the corresponding β-ketoesters. The optical purity was increased to >99% with one recrystallization [9]. A convenient and efficient synthesis of ligands used to prepare homogeneous chiral hydrogenation catalysts was described starting from the stereoselective hydrogenation of acetylacetone [10] (figure 1). The hydrogenation was developed by Izumi's group and commercialized by Wako Pure Chemicals Ind. [6,11].

Finally, Raney nickel modified with (R,R)-tartaric acid/NaBr was shown to be an effective catalyst for the asymmetric hydrogenation of an intermediate in the synthesis of tetrahydrolipstatin, a pancreatic lipase inhibitor (figure 2) developed by Hoffmann-LaRoche (100% chemical yield, ee 90-92%, 6-100 kg scale) [12].

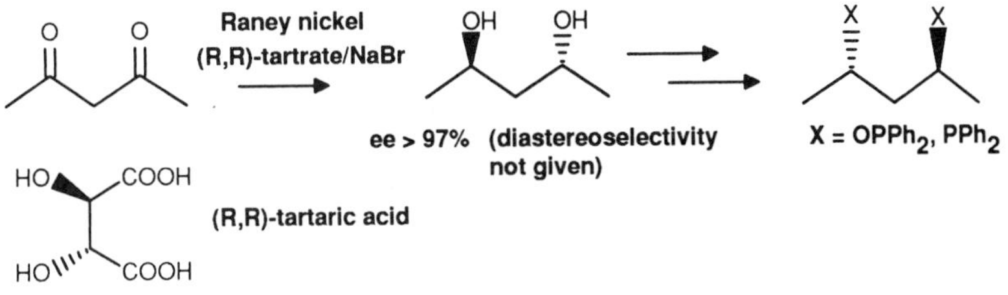

Figure 1. Synthesis of diphosphine ligands.

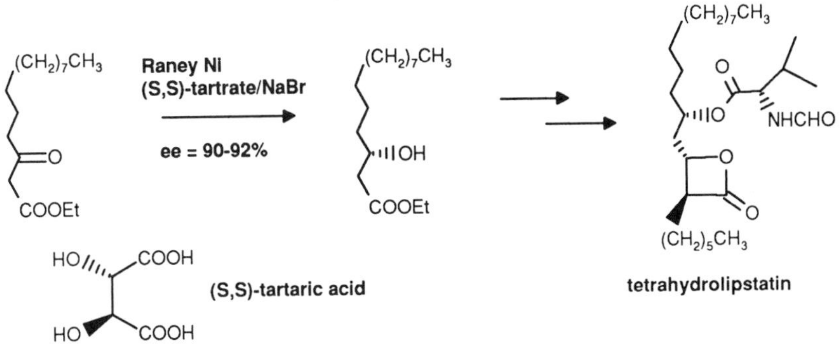

Figure 2. Synthesis of tetrahydrolipstatin.

Update on Recent Publications. Since the symposium contributions of Harada, Tai and Webb describe the use of modified Ni catalysts, recent publications are mentioned only summarily. Klabunovskii [13] developed correlations between the adsorption constants of aminoacids and optical yields. The conformation of the ternary complexes of different metals, modifiers and β-ketoesters was shown to correlate with the absolute configuration of the product hydroxyesters. Brunner et al. [14] described condensation of metals as a new method for preparing various tartrate/NaBr modified catalysts. Enantioselectivity for the hydrogenation of β-ketoesters and β-diketones was moderate. Keane and Webb [15-17] reported in great detail the effect of the modification procedure on the catalytic properties of Ni/SiO$_2$ catalysts using tartaric acid and aminoacids as modifiers. Tai and Harada [18] investigated ultra sonicated Raney nickel that is the most selective (ee's up to 94%) and active hydrogenation catalyst for β-ketoesters and β-diketones.

Cinchona Modified Pt and Pd Catalysts. The cinchona modified platinum catalysts are at the moment among the most selective catalytic systems known for the hydrogenation of α-ketoacid derivatives [4a]. The corresponding R-hydroxy derivatives are formed with good to very good optical yields while the S-enantiomers are usually obtained with somewhat lower enantioselectivities. At the present time the best catalysts are commercially available 5% Pt/Al$_2$O$_3$ catalysts with low dispersion and a rather large pore volume. 10,11-Dihydrocinchonidine (HCd) and O-methyl-HCd are the best modifiers [19]. Other inorganic carriers like SiO$_2$ or BaCO$_3$ or various carbon supports and different cinchona alkaloids are also suitable. The catalyst has to be pretreated in hydrogen at 300-400 °C before the reaction. The modifier can be directly added to the reaction solution but more elaborate modification procedures have also been reported [5].

Synthetic Applications. Two potential intermediates for the angiotensin-converting enzyme inhibitor benazepril were synthesized using cinchona modified Pt and Pd catalysts (figure 3). The hydrogenation of the α-ketoester has been developed and scaled-up into a production process (10-200 kg scale, chemical yield >98%, ee 79-82%) [20]. The novel Pd catalyzed enantioselective hydrodechlorination reaction is a potential alternative to the established synthesis where the racemic a-bromobenzazepinone is used [21]. At the moment, both selectivity and productivity of the Pd-cinchona catalyst are too low for an efficient synthesis.

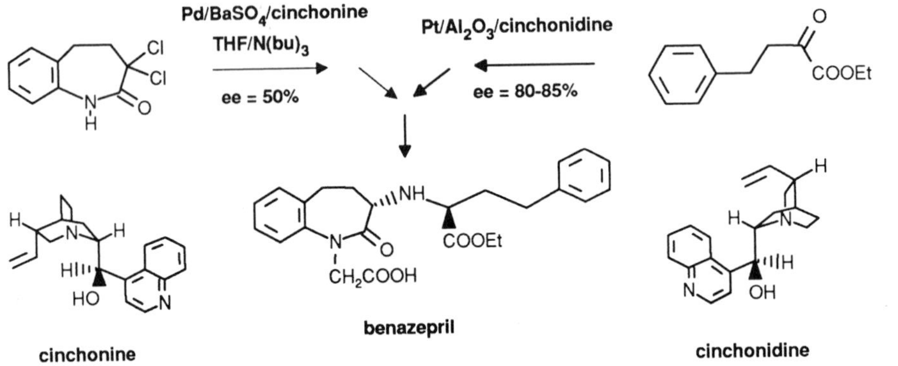

Figure 3. Synthesis of benazepril.

Update on Recent Publications. Several research groups are at the moment actively involved in the study of cinchona modified catalysts. Recently published papers deal with mechanistic investigations, the search for new catalyst types and new transformations.

Mechanistic investigations are concentrating on finding an explanation for the remarkable rate enhancement that is associated with the asymmetric hydrogenation of ethyl pyruvate using Pt catalysts modified with cinchonidine derivatives. Wells and coworkers [22] explain the observed rate enhancement with three effects: i) the activation of the substrate by the quinuclidine nitrogen (from the effect of quinuclidine addition); ii) a higher hydrogen coverage of the modified catalyst (enhanced H/D exchange); iii) the electronic effect of the adsorbed quinoline part (from the effect of quinoline addition). Several studies report investigations on the effect of modifier concentrations on rate and optical yield. At low cinchona concentrations, the results of Wells et al. [23,24] for EUROPT-1 (see figure 4) and of Blaser et al. [25] for Pt/Al$_2$O$_3$ (see figure 5) at high H$_2$ pressures are quite similar. Augustine [26] described a more complex behavior for Pt/Al$_2$O$_3$ at low pressure (see figure 6). Especially the change of the absolute configuration of the produced ethyl lactate at extremely low modifier concentration is puzzling. Augustine suggested different types of surface sites. At high modifier concentration both Blaser and Augustine observe in many cases a decrease of the optical yield. The results of Blaser can be modeled for low and high modifier concentrations using a two and three site model, respectively, as shown in figure 5 [25]. The results of Wells et al. [23,24] can also be described by such a model (see figure 4) while this is not possible for the results described by Augustine.

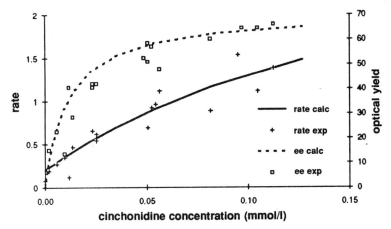

Figure 4. Effect of the modifier concentration on rate and optical yield for the hydrogenation of methyl pyruvate (EUROPT-1; cinchonidine; EtOH; 10 bar). The experimental points were taken from [24], the calculated curves were obtained using the two site model described in [25], assuming rates of 0.2 mol/g.h and of 3.0 mol/g.h for unmodified and modified catalyst, respectively, and an enantioselectivity of 70% for the modified sites.

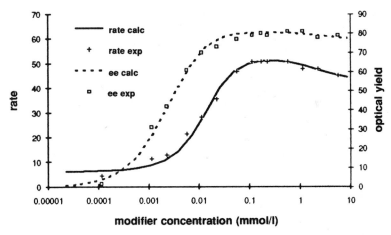

Figure 5. Effect of modifier concentration on rate and optical yield for the hydrogenation of ethyl pyruvate (Pt/Al$_2$O$_3$; dihydrocinchonidine; toluene; 20 bar) [25].

Margitfalvi and Baiker carefully investigated the behavior of cinchonidine in the reaction mixture [27] and the fate of the modifier during the reaction [28]. They find that cinchonidine is hydrogenated very fast to 10,11-dihydrocinchonidine, which is then hydrogenated further more slowly. Wells and coworkers [29] described the effect of mixtures of modifiers. Baiker et al. [30] published preliminary molecular modeling studies of the interactions of cinchonidine and its protonated form with ethyl pyruvate.

The mechanistic picture that emerges from these studies is far from uniform. Most investigators probably agree that rate enhancement and enantiocontrol are closely connected effects and should be discussed together. Or, to express the problem differently, one must explain why only the formation of one enantiomer is accelerated as shown in figure 7. There is also agreement that very specific interactions between modifier and substrate are necessary in order to give such high enantioselectivities. There is much less agreement as to where and how these interactions take place: Is one cinchona molecule involved in the rate and product controlling step [25,26,30] or is it an array of cinchona molecules [22-24,29]? How important are the interactions between modifier and α-ketoester in solution [27,28,31]?

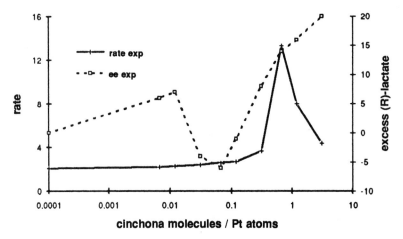

Figure 6. Effect of modifier concentration on rate and optical yield for the hydrogenation of ethyl pyruvate (Pt/Al$_2$O$_3$; dihydrocinchonidine; methyl acetate; 1 bar), adapted from [26].

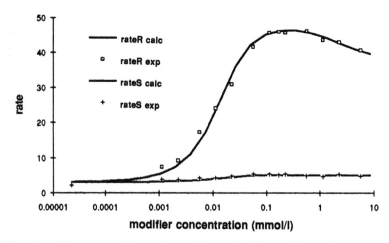

Figure 7. Effect of modifier concentration on the rate of formation of R- and S-lactate for the hydrogenation of ethyl pyruvate (Pt/Al$_2$O$_3$; dihydrocinchonidine; toluene; 20 bar) [25].

Two new types of Pt-cinchona systems have been described. Blaser and Müller [32] reported the immobilization of a cinchonidine derivative on the silica surface of a Pt/SiO$_2$ catalyst (see figure 8). They demonstrated that the attached modifier is indeed effective and leads to rates and ee's that are comparable to the normally modified catalyst. Unfortunately, the catalyst was not re-usable. Bhaduri et al. [33] reported the preparation of a completely different catalyst. They anchored anionic Pt and Ru carbonyl clusters on a 20% cross-linked polystyrene with pendant quaternary cinchona alkaloid functionalities (see figure 8). After heating to 80 °C, the catalysts were active and enantioselective for the hydrogenation of different ketoesters. Best results were obtained with a Pt-cinchonidine catalyst for the hydrogenation of methyl pyruvate with ee's up to 80%. These results are surprising for several reasons. On the one hand, structure-activity studies [19] on Pt/Al_2O_3 catalysts have shown that small Pt crystallites are unselective and that quaternization of the quinuclidine nitrogen leads to completely unselective modifiers. On the other hand, it was found that cinchona modified Ru/Al_2O_3 catalysts give a racemic hydrogenation [34]. This probably means that these new cluster catalysts must have a quite different mode of action than the modified supported types used until now.

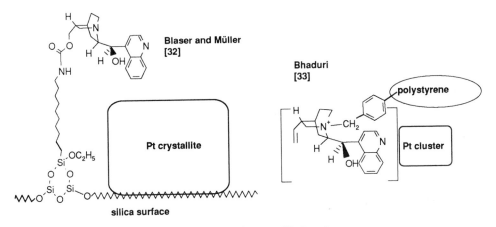

Figure 8. New cinchona modified catalyst.

Concerning new transformations, one important limitation of modified hydrogenation catalysts is their rather high specificity i.e. only few types of substrates are transformed with good enantioselectivity [4]. Two recent publications slightly extend the use of the cinchona Pt system. Wells et al. [35] describe the chemo- and enantioselective hydrogenation of α-diketones to the corresponding α-hydroxyketones in ee's up to 38%. Blaser and Jalett [36] report on the hydrogenation of α-ketoacids with enantioselectivities of over 80%. In both cases, the absolute configuration of the hydroxy compounds is the same as the one observed for the α-ketoester for the same catalytic system and there is also a similar rate enhancement.

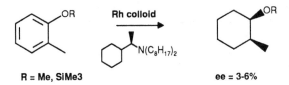

Figure 9. Enantioselective hydrogenation of aromatic compounds.

New Catalytic Systems. As already mentioned, it is very difficult to find or develop new chiral catalysts. The report by Lemaire et al. [37] on Rh colloids stabilized and modified by chiral amines is therefore of interest (figure 9). The optical yields observed for the enantioselective hydrogenation of substituted aromatic rings are low but significant. These results give some hope that the asymmetric hydrogenation of aromatic systems might be feasible.

Modified Metal Oxides

Solid acids and bases are increasingly used for the catalytic synthesis of fine chemicals. It is therefore quite natural that attempts have been made to render these materials chiral and to carry out reactions enantioselectively. Generally, this is done by modification with a chiral compound but an example where crystalline metal tartrates were used is also known [38]. With the exception of the Ti pillared clays [39,40], the catalysts are not yet practically useful. However, some of devised concepts are interesting.

Zeolites and Clays. In principle, it is a fascinating idea to use a well-defined zeolite to control the stereochemistry of a given reaction either by synthesizing a chiral zeolite or by inserting a chiral modifier into the cavities. First attempts were described in two patents by Dessau [41] who contacted Pt- or Rh-loaded H-ZSM-5 with (S)-2-phenyl-ethyl amine and used this catalyst for hydrogenation and hydroformylation reactions. No details with regard to optical yields or catalyst structure were disclosed. Very recently, it was demonstrated for the first time that a chiral zeolite can indeed perform enantio-discrimination : zeolite β, partially enriched in polymorph A, catalyzed the ring opening of an epoxide with low but significant 5% ee [42]. Zeolites have also been used as supports for the immobilization of metal complexes (see section "Immobilized Metal Complexes").

Titanium-pillared montmorillonite modified with tartrates are very selective hetero-geneous catalysts for the Sharpless epoxidation [39] as well as for the oxidation of aromatic sulfides [40]. Vanadium pillared clays were less effective for the sulfide oxidation [43]. At this time it is not clear whether the preparation of these unusual catalysts is reproducible without difficulty. A heterogeneous version with such good selectivity would be of tremendous advantage, since the Sharpless epoxidation is widely used in preparative chemistry and also for two industrial applications [44].

Modified Amorphous Metal Oxides. Morihara described the creation of so-called "molecular footprints" on the surface of an Al-doped silica gel surface using an aminoacid derivative as chiral template molecule [45]. After removal of the template, the catalyst showed low but significant enantioselectivity for the hydrolysis of a structurally related anhydride. It was shown convincingly that the template molecule is not responsible for this effect. Moriguchi [46] used a more classical strategy to modify different metal oxides with histidine which then showed a modest enantioselectivity for the solvolysis of activated aminoacid esters. Cativiela and coworkers [47] treated silica or alumina with diethyl-aluminiumchloride and menthol. The resulting modified material catalyzed the Diels-Alder reaction between cylopentadiene and methacrolein with modest enantioselectivity (ee ca. 25-30%).

All these catalysts are not yet practically important but rather demonstrate that amorphous metal oxides can be modified successfully.

CHIRAL POLYMERS AS CATALYSTS OR SUPPORTS

A different strategy for preparing enantioselective heterogeneous catalysts is the application of natural or artificial chiral polymers. In some cases, the resulting catalyst might not be heterogeneous in the strictest sense. Indeed, historically these materials were investigated because of the interest in understanding enzymes [4a]. The first hydrogenation catalyst with high enantioselectivity was a Pd/silk fibroin catalyst where ee's up to 66% were reported in 1956 by Akabori and Isoda for the hydrogenation of an oxazolinone derivative [4a]. However, it was very difficult to reproduce the results and other approaches were pursued. In the mean time, some progress has been made even though only very few successful catalysts of this type are known. This is probably again due to the difficulty to design such materials that can control the stereochemistry of a reaction. Whereas some synthetic applications have been described, the technical feasibility of these catalysts has not yet been demonstrated.

Chiral Polymers as Catalysts

The most useful and selective polymer-based catalysts at this time are synthetic poly(aminoacid) derivatives used for the catalytic epoxidation of electron deficient olefins with H_2O_2 in a two-phase reaction system [48]. Several parameters were shown to be of importance for the catalytic performance: the type of aminoacid, the degree of polymerization, the substituent at the terminal amino group and the organic solvent. From these and other observations it was inferred that supramolecular interactions inside the polymer aggregate are important for catalysis and stereocontrol. Itsuno et al. [49] described the preparation of poly(aminoacids) grafted onto 2% cross-linked polystyrene. The resulting catalysts were highly selective for the epoxidation of several substituted chalcones with optical yields up to 99%, re-use was possible. Boulahia et al. [50] used acrylamide-methacrylamide copolymers as supports and reported optical yields of ca. 80% and re-use was problematic.

Poly(aminoacid) catalysts were used to prepare chiral flavonoid intermediates with medium enantiomeric excess [51] and intermediates for a class of leukotriene receptor antagonists with very high optical yields [52] (figure 10). N-Arylmethylchitosan, a poly(aminosugar), was described as active catalyst for the addition of KCN and acetanhydride to imines, leading to N-acetylamino nitrile derivatives in optical yields of up to 60% [53].

In principle, it is an interesting idea to use a protein as a macromolecular chiral catalyst or template. Promising results were reported for several transformations using Bovine Serum Albumin (molecular weight 60-90 000) as catalyst. Optical yields between 62% and 85% (for a sulfide oxidation) were reported. However, these reactions were carried out in homogeneous phase [54-58].

Chiral Polymers as Supports

As mentioned above, silk fibroin was used successfully as a chiral support. Other materials such as cellulose or polysaccharides were ineffective [4a]. The use of a polystyrene with pendant quaternary cinchona alkaloid functionalities as support for Pt and Ru carbonyl clusters is described above in section "Cinchona Modified Pt and Pd Catalysts : Update on Recent Publications". Some recent results (ee's 44-70%) indicate that such an approach has some potential when metal complexes are attached to an enzyme or a synthetic chiral polymer [59-62], but again, these catalysts are homogeneous.

Liquid Crystals and Gels

Liquid crystals and gels could be regarded as labile organic polymers. The conditions inside such a material are different from those in solution. At the moment, the synthetic application of such systems does not seem very attractive but this might change in the future. We will present a short summary on some recent results.

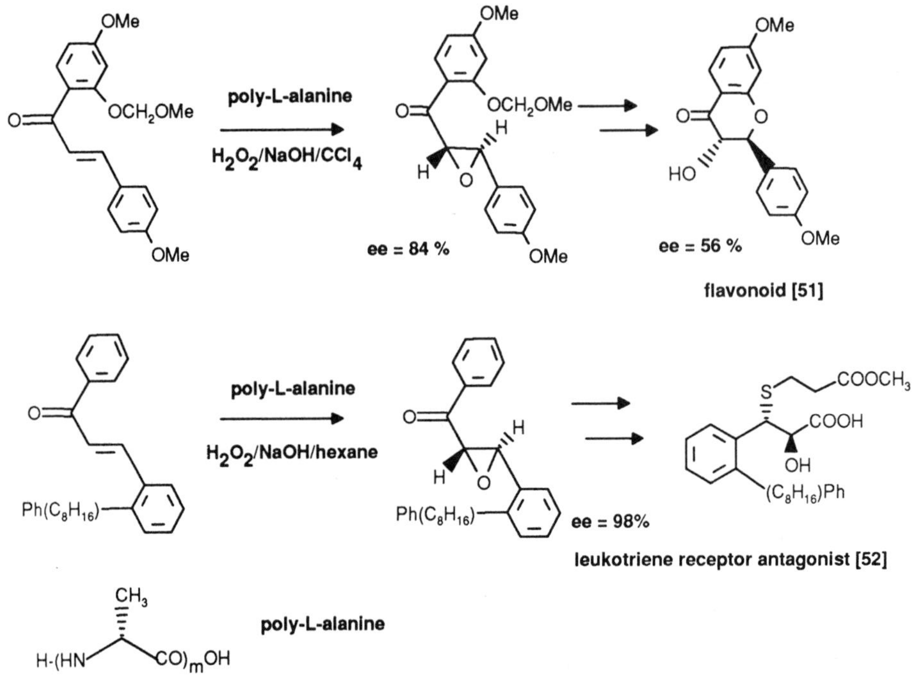

Figure 10. Synthetic applications of poly(aminoacid) catalysts.

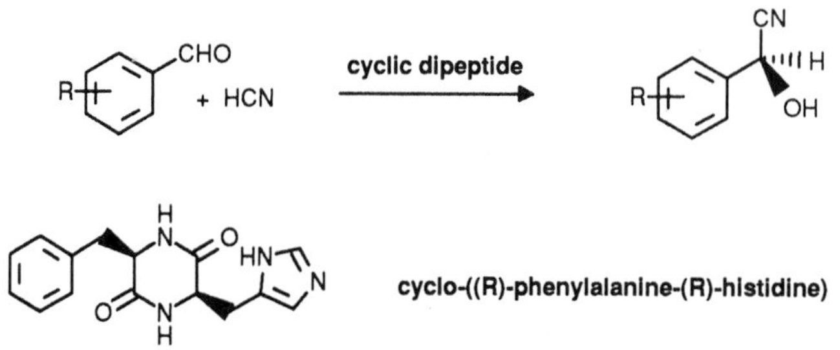

Figure 11. Hydrocyanation of an aldehyde catalyzed by a cyclic dipeptide.

In preliminary experiments, cholesteric liquid crystals were utilized for the decarboxylation of malonic acid derivatives (best ee 18%) [63], for the equilibration of oxaziridines (best ee 20%) [64] and the hydrogenation of enamides with Wilkinson's catalyst (best ee 16%) [65]. Even though the optical yields are modest, they are still better than when chiral solvents are used. This is explained by the higher order in liquid crystals [63].

Cyclic dipeptides are efficient catalysts for the hydrocyanation of aromatic aldehydes if the reaction is carried out in a "clear gel" (best ee 92%) [66] (figure 11). Even higher optical yields are claimed when the catalyst is adsorbed on a non-ionic polymer resin (ee >98%)

[67]. These findings indicate that there is a higher order both in the gel and in the adsorbed state leading to better enantiocontrol compared to the homogeneous catalyst.

IMMOBILIZED METAL COMPLEXES

In many cases, homogeneous catalysts have activity and selectivity patterns that can not be obtained with classical heterogeneous catalysts. The combined efforts of numerous research groups both at universities and in industry have resulted in a large number of enantioselective soluble catalysts for many different chemical transformations. As already mentioned, their separation and handling properties are often problematic. On the other hand, the technical use of heterogeneous catalysts is well established. The most promising strategy to combine the best properties of the two catalyst types is the heterogenization or immobilization of active metal complexes on insoluble supports or carriers [68-71]. Unfortunately, the attachment of a homogeneous catalyst on a support often leads to a change in its catalytic properties. Most of the time the consequences are negative but sometimes the performance is improved. Even though these effects are poorly understood and ill defined, several factors may be distinguished. i) Interactions between functional groups on the surface of the support and the metal center can affect the catalytic performance. The introduction of additional chiral groups has been used to influence the enantioselectivity as demonstrated by Stille [72-74]. ii) Restricted conformational flexibility through geometrical confinement can be positive as described by Corma et al. [75-77] for proline amide Rh complexes negative as observed by Pugin and Müller [78] for a diphosphine Rh complex. iii) Attaching a complex to a rigid support can lead to so-called site isolation, i.e. different active centers no longer interact with each other. The most remarkable positive effects are observed for complexes that are prone to form inactive dimers [79].

In this chapter, different methods will be outlined for producing immobilized analogs of active homogeneous catalysts that have a potential use in the synthesis of fine chemicals. Their most likely application will be for the synthesis of expensive, low volume products with delicate selectivity problems but, to our knowledge, none is applied on a technical scale.

Requirements for Practically Useful Immobilized Catalysts

The immobilization of enantioselective catalysts is still a young field. So far, most of the work was done to show the feasibility of the immobilization concepts. Based on the published work and our own experience, we think that several requirements should be met in order to make immobilized enantioselective catalysts practically useful.

Preparation Methods
- Generally applicable. It is not yet possible to predict which ligand and support will be suitable for a given substrate and process. Therefore, flexible methods that allow different combinations of support, linker and ligand are preferred.
- Simple and efficient. The cost of immobilization has to be lower than the cost for the separation of the homogeneous catalyst.
- Reasonable "molecular weight". For practical reasons, the "molecular weight" of a catalyst immobilized to a support should not exceed 10 kD per mole of active sites. Therefore, high loading densities should be possible.

Catalytic Properties and Handling
- Catalytic performance. The performance (selectivity, activity, productivity) of an immobilized catalyst should be comparable or better than that of the corresponding free catalysts.
- Separation. Separation should be achieved by a simple filtration and at least 95% of the catalyst should be recovered.
- Metal leaching. The metal complex must be stable enough because very often the metal is expensive and the limit for metal contamination in biologically active compounds is very low.
- Re-use. Though not mandatory, this would be a great advantage from an economic point of view.

Supports
- Chemical stability. The support should be inert towards all reagents under the reaction conditions.
- Mechanical stability. The support should not change particle size when the reaction mixture is stirred.
- Separation. Simple filtration is preferable, separating soluble polymer supports requires costly methods like ultrafiltration or precipitation.
- Solvent compatibility. The support should function in different solvents, otherwise this will restrict its application.
- Availability. The support material must be commercially available in reproducible quality (texture, purity etc.).

In the next sections we shall discuss the preparation methods and catalytic properties of immobilized catalysts in more detail and asses them on the basis of these requirements.

Heterogenization via a Covalently Bound Ligand

Attaching the ligands via a suitable bifunctional linker or tether to the support is by far the most important and versatile strategy for the heterogenization of metal complexes. In order to avoid metal leaching, bidentate phosphine or amine ligands are used in most cases. The metal complexes are usually prepared after the immobilization of the ligand. So far, most work has been carried out with Rh and Pt diphosphine catalysts for enantioselective hydrogenation and hydroformylation reactions.

Support Materials. Many organic polymers and inorganic solids have been used for the covalent immobilization of metal complex catalysts (see figure 12). There are many cases which show that the support properties (e.g. chemical composition, degree of crosslinking, porosity etc.) influence the catalytic performance of immobilized metal complexes. Kagan [80] found for example that a Rh diphosphine catalyst bound to 2% crosslinked polystyrene hydrogenated apolar substrates in benzene but was not active in an alcoholic solution, because the polystyrene support does not swell in polar solvents.

Linear polymers. Non-crosslinked, linear polymers are soluble in most solvents. The immobilization of catalysts on such supports will not induce significant changes in the mobility and the ligand sphere of the metal complex. Therefore, the catalytic properties will remain practically unchanged. However, the separation of such catalysts is often problematic and costly since it is done either by ultrafiltration or by precipitation.

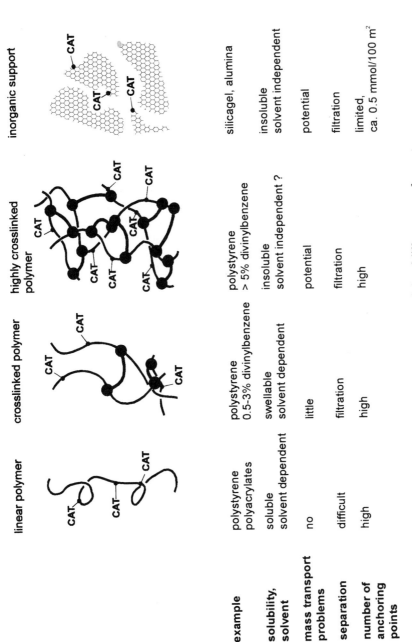

Figure 12. Schematic representation and characteristics of different types of supports.

Crosslinked polymers. Crosslinked polymers must be regarded as 3-dimensional networks. In suitable solvents they swell and the reactant molecules have access to the inner surface. As long as the mesh size is large enough, the metal complex immobilized inside such a support does not experience important changes. If, however, the degree of crosslinking is increased or the wrong solvent is used, the mesh size decreases. As a result, interactions between the polymer-network and the ligand sphere of the metal are stronger and, in addition, mass transport problems can occur. The result is usually a decreased activity and selectivity of the catalyst.

Slightly crosslinked polymers are widely used in solid state peptide synthesis and a large variety is commercially available. They can easily be separated by filtration or by decantation. Highly crosslinked polymers are practically not swellable. To guarantee a sufficient mass transport these supports must be porous. Also, they should have a large specific surface area for binding the catalyst molecules. Many macroporous polymers with these characteristics have been developed as adsorbing agents. They are commercially available in large quantities and are suitable as catalyst supports.

Inorganic supports (metal oxides). These insoluble supports must have a large specific surface area like the highly crosslinked polymers. Ligands bearing a triethoxysilane function are immobilized via surface hydroxy groups. Based on geometric arguments, the maximum possible loading for a typical diphosphine ligand is estimated to be about 0.7 mmol / m^2 [78]. To obtain supported catalysts with a reasonable weight per mole of active sites (< 10 kD) the specific surface area therefore has to be at least 150 m^2/g. To avoid mass transport problems, the pore-size should be considerably larger than the size of the substrate. The most frequently used inorganic supports are silica gels. Different types covering a large range of specific surface areas and pore sizes are commercially available. As an alternative, zeolites or Aerosols can be used. The latter are non-porous SiO_2-spheres typically in the range of 10-30 nm. However, due to their small size they are more difficult to handle than silica gels.

Immobilization Methods. Several immobilization strategies have been shown to give stable and active heterogeneous catalysts. The three most useful ones are discussed in the following paragraphs (see figure 13).

Grafting i.e. reacting a functionalized ligand or metal complex with reactive groups of an organic or inorganic support. This method has several advantages. Different suitable supports are commercially available (see above) and many methods for introducing reactive groups onto non-functionalized polymers are described in the literature [69,81,82]. Their properties (e.g. solubility or swellabillity, particle size, separation, purity) can be controlled and checked before use. Also, it is possible to purify, characterize and test the ligands before immobilization. Finally, the ligands will preferentially bind at locations that are also accessible for the substrate during the catalytic reaction. Grafting is probably the method that can most easily be controlled and scaled up.

Grafting ligands to inorganic supports requires ligands that are functionalized with a trialkoxysilane group, that readily reacts with the surface OH-groups (see figure 14). It is assumed, that a stable link is formed with one or two surface oxygen atoms [83]. There are different ways of functionalizing a ligand with a trialkoxysilane group. Cerny [84] designed a synthesis that is specific for the DIOP-ligand. Nagel [85] and Pugin [78] used different types of linkers (drawn bold in figure 14) that can be attached to any ligand with an NH- or OH-group. At present, the most efficient method is the use of commercially available trialkoxy-isocyanate linkers [78]. Due to their high reactivity, quantitative yields are obtained under mild conditions.

a) grafting

b) solid state synthesis

c) copolymerization

Figure 13. Immobilization strategies for covalently bound complexes.

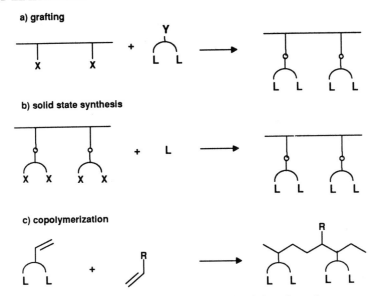

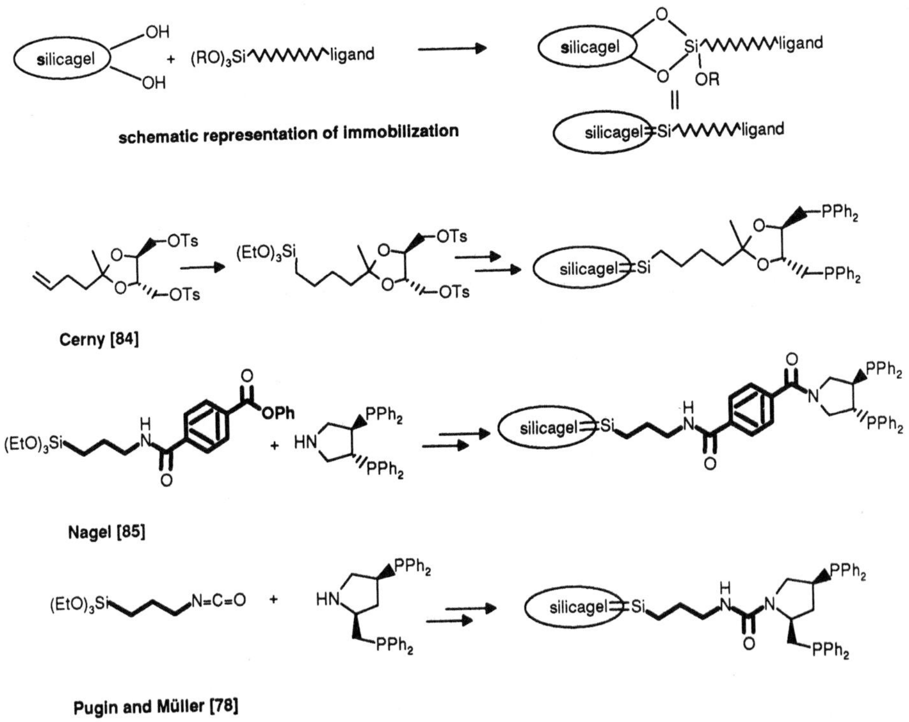

schematic representation of immobilization

Cerny [84]

Nagel [85]

Pugin and Müller [78]

Figure 14. Representative examples for catalysts grafted on inorganic supports.

The grafting method has also successfully been applied to bind amine or oxygen containing ligands onto functionalized polystyrenes [see e.g. 86,87] (see figure 15).

Interestingly, no chiral diphosphine ligands have been attached to organic polymers, maybe as a consequence of a negative statement by Stille [88] who favored the copolymerization method.

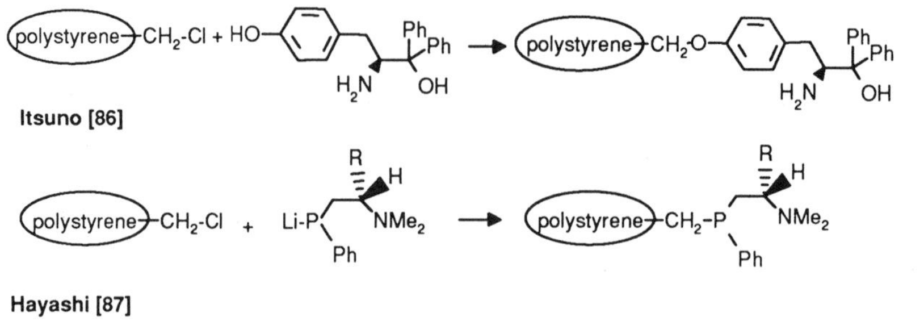

Figure 15. Representative examples for catalysts grafted on organic polymers.

Solid state synthesis. Here, the ligand is assembled on the polymer. (figure 16). This method is seldom used but there are a few examples in the literature [e.g. 73,80,89]. There is a potential risk to produce defective ligands due to incomplete conversion possibly leading to reduced enantioselectivity of the catalyst.

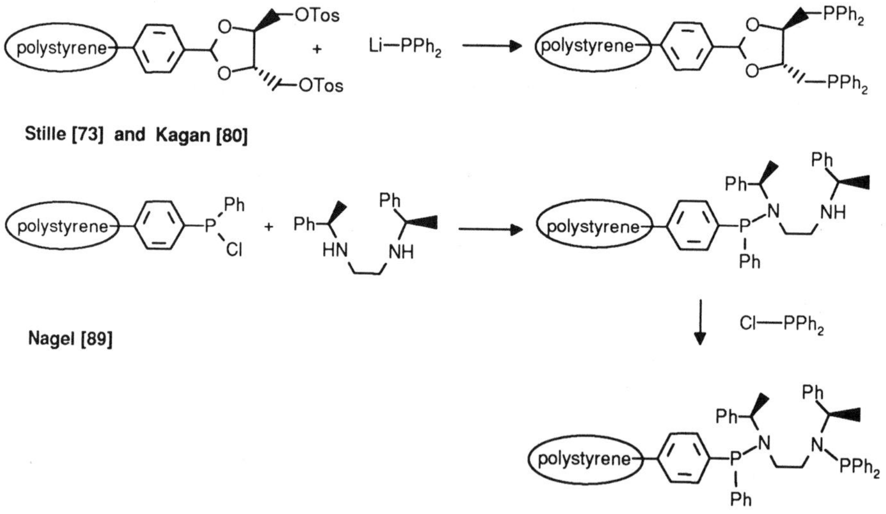

Figure 16. Catalysts prepared by solid state synthesis.

Copolymerization of ligands functionalized with an (activated) olefinic group with a suitable monomer (figure 17). This approach has been used extensively but has several potential problems [90-93]. It is difficult to predict and control the properties of the resulting polymer that is a random sequence of the original monomer units. In addition, the formation of inaccessible immobilized ligands cannot be excluded. Finally, unwanted polymerization could occur during the synthesis or the storage of the functionalized ligand.

The copolymerized diphosphine complexes were mainly tested in hydrogenation reactions. Generally, good optical yields were obtained, but activities were generally low, possibly due to mass transport restrictions or inaccessible complexes.

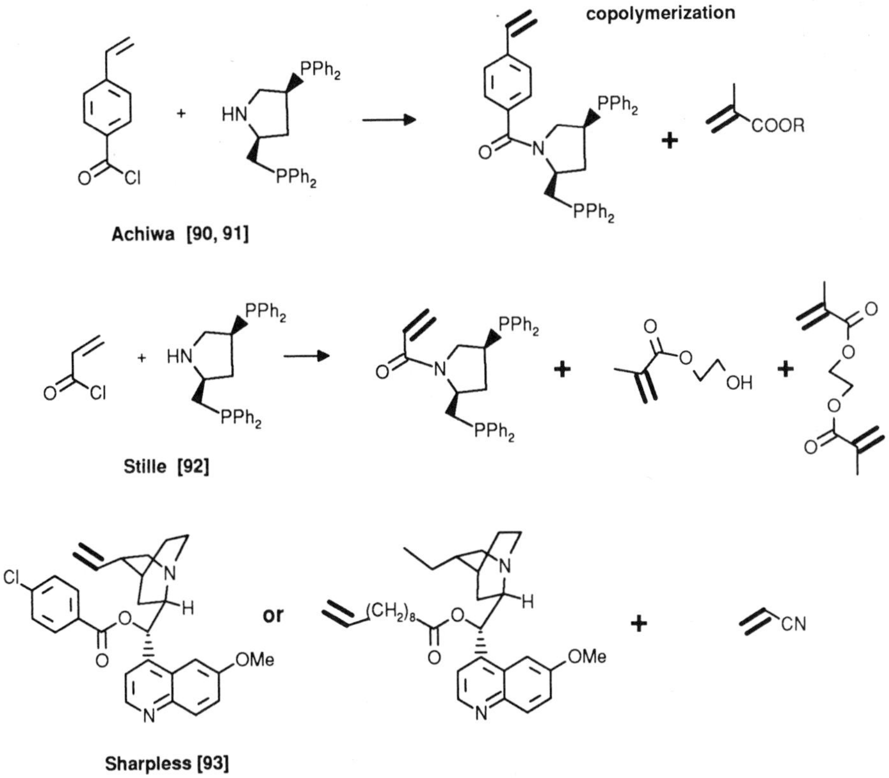

Figure 17. Representative examples for copolymerized catalysts.

Heterogenization via Adsorption and Ion Pair Formation

This approach relies on various adsorptive interactions between a carrier and a metal complex. The advantage of this approach is the easy preparation of the heterogenized catalyst by a simple adsorption procedure very often without the need to functionalize the ligand. Electrostatic forces were used to bind cationic Rh diphosphine complexes to anionic resins [68,94]. The resulting catalysts could be recycled 20 times with very little leaching. Toth et al. [95] functionalized diphosphine ligands with amino groups and adsorbed the corresponding Rh complexes on Nafion. Another approach by Inoue relied on the interaction of lipophilic ligands with surface methylated silica [96]. Further examples were described by Brunner [97] and Inoue [98]. Depending on the method used, strong solvent effects have to be expected and this might limit the use of this elegant immobilization strategy. With few exceptions [68], adsorbed complexes have lower enantioselectivities than covalently bound catalysts.

Heterogenization via Entrapment

This method relies on the size of the metal complex rather than on a specific adsorptive interaction. Various materials with small regular pores, especially zeolites, have been used for the entrapment of metallic species, but the only example of a chiral catalyst is a Rh diphosphine complex entrapped in the interlayers of Smectite [99]. The resulting catalyst was active for the enantioselective hydrogenation of N-acetamido acrylic acid (ee 75%).

USEFUL HETEROGENEOUS ENANTIOSELECTIVE CATALYSTS

In this section we give an overview on chiral heterogeneous catalyst systems that we consider to be useful for synthetic purposes, i.e., they fulfill most of the requirements defined above. Our selection is not comprehensive since sometimes practically the same catalytic systems have been described by different groups. The inclusion of a catalyst in our tables does of course not mean that it can be applied routinely like any organic reagent.

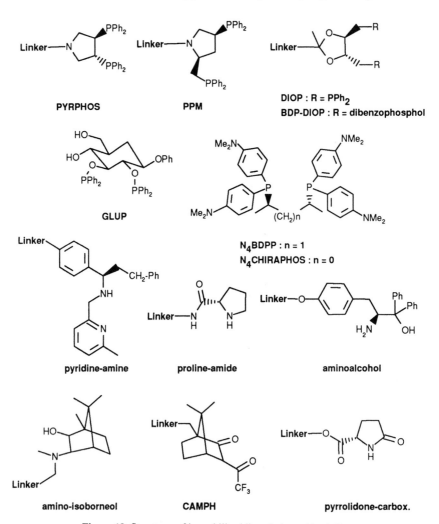

Figure 18. Structures of immobilized ligands (see tables 1-3).

Table 1. Potentially useful heterogeneous catalysts for enantioselective hydrogenation and reduction.

Substrate, reagents	metal	modifier/ligand	support	ee (%)	TOF(hr⁻¹)	p(bar) / T(°C)	re-use	Fig.	Ref.
hydrogenations, modified heterogeneous catalysts									
β-ketoesters	Raney-Ni	tartrate/NaBr	-	94	1.5	100/100	(yes)	2	[18]
β-ketoesters	Raney-Ni	tartrate/NaBr	-	91	0.6	100/100	(yes)	1	[18]
β-diketones	Pt	cinchona deriv.	Al_2O_3	95	>50'000	70/25	(yes)	3	[4b]
α-ketoesters	Pt	cinchona deriv.	Al_2O_3	85	>1000	100/25	?	3	[4b]
α-ketoacids									
hydrogenations, immobilized catalysts									
enamides	Rh	PYRPHOS	Silicagel	100	500	40/25	yes	18	[85]
enamides	Rh,	Proline-amide	USY-Zeolite	99	10	5/65	yes	18	[75, 76]
enamides	Rh,	Proline-amide	Silicagel	92	8	5/65	yes	18	[75, 76]
enamides	Rh	PPM, PYRPHOS, DIOP	Silicagel	94	>670	1/25	yes	18	[78]
enamides	Rh	PPM, PYRPHOS, DIOP	Silicagel	63	30	40/30	yes	18	[78]
N-phenyl-imines	Ir	PPM	HEMAa)	91	low	50/25	(yes)	18	[92]
enamides	Rh	DIOP	PMVK (2%)b)	76	25	1/25	(yes)	18	[72]
pantolactone	Rh	PPM	HEMAa)	73	2	50/50	(yes)	18	[90]
hydrogenations, adsorbed catalysts									
enamides	Rh	GLUP	sulfonated PSc)	95	ca. 600	1/25	yes	18	[68]
enamides	Rh	GLUP	silicagel / Ph-SO₃H	95	ca. 500	1/25	yes	18	[94]
enamides	Rh	DIOP, PPM	reversed phase silica	89	?	20/50	(yes)	18	[96]
enamides	Rh	N₄BDPP, N₄CHIRAPHOS	sulfonated PSc)	80	<600	14/20	yes	28	[95]
reductions, immobilized catalysts									
ketones + iPrOH	Ir	pyridine-amine	polymethacrylate	84	180	-/60	?	18	[100]
O-methyloximes + BH₃	B	aminoalcohol	PS(30%)c)	99	?	?	yes	18	[105]
ketones + BH₃	B	aminoalcohol	PS(2%)c)	93	ca. 3d)	-/25	yes	18	[106]

a) 2-hydroxy-methacrylate; b) polymethylvinylketone(2% crosslinked); c) polystyrene(% crosslinked); d) continuous flow system

Table 2. Potentially useful heterogeneous catalysts for enantioselective C-C bond formation.

Substrate, reagents	metal	modifier	support	ee (%)	TOF(h^{-1})	re-use	Fig.	Ref.
addition to C=C-C=O								
alkenones + R_2Zn	Ni	proline-amide	USY-zeolite	95	1	?	18	[77]
addition to C=O								
aldehydes + R_2Zn	Zn	amino-alcohol	PS (20%)[a]	99	3	yes	18	[86]
aldehydes + R_2Zn	Zn	amino-alcohol	PS(2%)[a]	95	0.8	yes	18	[107]
aldehydes + HCN	-	dipeptide	PS	98	20	yes	11	[67]
hydroformylation								
styrene / triethyl-orthoformate	Pt/SnCl$_2$	PPM	PS(10%)[a]	98	0.6	?	18	[101]
styrenes	Pt/SnCl$_2$	BDP-DIOP	PS(0-10%)[a]	56-65	11	yes	18	[108]
cyclopropanation								
styrene + diazodimedone	Cu	CAMPH	silica	98	?	(yes)	18	[109]
methylbutenyl diazoacetate	Rh	pyrrolidone-carbox.	polyethylene	98	?	(yes)	18	[110]

a) polystyrene, 0-20% crosslinked with divinylbenzene

Table 3. Potentially useful heterogeneous catalysts for enantioselective oxidation.

Substrate and reagents	metal	modifier	support	ee (%)	TOF(h^{-1})	T(°C)	re-use	Fig.	ref.
epoxidation									
allylic alcohols + tBuOOH	Ti	dialkyl tartrate	PILC[a]	98	8	-15	(no)		[39]
aryl sulfides + tBuOOH	Ti	dialkyl tartrate	PILC[a]	92	9	-20	(yes)		[40]
chalcones + H_2O_2	-	poly(amino acid)	PS,2%[b]	99	0.1	25	yes	10	[49]
chalcones + H_2O_2	-	poly(amino acid)	-	98	0.1	25	yes	10	[52]
dihydroxylation									
alkenes + NMO[d]	Os	quinidines	PAN[c]	93	5	10	yes	17	[93]

a) pillared clays (montmorillonite); b) polystyrene, 2% cross linked; c) polyacrynitril; d) N-methylmorpholine-N-oxide

Especially the preparation and characterization of some of these solid organic or inorganic materials will be difficult for many organic chemists. Almost none of the catalysts described above is available commercially. Therefore, in most cases a certain experience has to be acquired before heterogeneous catalysts of this complexity are used efficiently.

Some technical details concerning tables 1, 2 and 3: the enantiomeric excess (ee) is the best value reported in the literature. Turnover frequencies (TOF) are average values estimated from reaction times and catalyst concentrations (for metallic catalysts, the TOF's are based on total metal content). Re-use "yes" means that no loss in activity or ee was observed, "(yes)" that a significant decrease was reported. The structures of immobilized ligands are depicted in figure 18.

Hydrogenation and Reduction

Except for the modified heterogeneous catalysts, most of the results reported in table 1 have been obtained with standard test substrates, mainly for acetamido derivatives of cinnamic and acrylic acid. Up to now, the activity of all immobilized complexes is lower than for the homogeneous analogs that have TOF's 1000-2000 (1 bar H_2), whereas the optical yields are comparable. The Ir catalysts described by Kaschig [100] are exceptional because both activity and enantioselectivity is improved significantly by immobilization (ee 52%, TOF 90 h^{-1} for the homogeneous complex). Usually, one can expect that the catalytic performance of an immobilized catalyst will be comparable to the homogeneous analog when applied to other hydrogenation problems. Nevertheless, all catalyst parameters must be optimized for the desired transformation.

C-C-Bond Forming Reactions

All of the successful C-C bond forming transformations were carried out with model substrates like benzaldehyde or styrene (table 2). Their practical application in "real" syntheses could be more problematic than for the hydrogenation reactions because compatibility with other functional groups has seldom been demonstrated. For the addition reactions of dialkyl Zn, the activity for the immobilized system is lower than for the homogeneous analog but ee's are high and at least for one case, a continuous flow reactor was developed [86]. Even though much effort was put into the hydroformylation reaction, its practical application is problematic due to difficulties with the regioselectivity and to racemization of the product [101,102].

Oxidation Reactions

Both the enantioselective epoxidation and the dihydroxylation reactions are applied widely for different synthetic applications table 3). The homogeneous version of the Sharpless epoxidation occurs with slightly lower optical yields but with higher TOF (ca. 20-30 h^{-1}) [103]. The oxidation of aryl sulfides using soluble Ti-tartrate complexes is not catalytic. The most recent dihydroxylation catalysts of Sharpless [104] have not yet been immobilized.

CONCLUSIONS

After a period of stagnation, the fascinating area of heterogeneous enantioselective catalysis has again become a fashionable topic. Several research groups both in industry and at universities are now actively involved. Three general strategies are pursued: the chiral modification of metallic and oxidic heterogeneous catalysts, the application of chiral polymers and the immobilization of homogeneous catalysts. The situation for the three fields of research are quite different. Modified catalysts and chiral polymers have already been applied for the synthesis of active compounds or have even been developed for technical applications. However, their scope is still very narrow, i.e., only few transformations are catalyzed with useful optical yields. Immobilized catalysts on the other hand have been applied only for model reactions. Relatively few of the highly selective ligands have been immobilized but in principle, this is possible for many more. The potential scope of this approach is impressive because numerous catalytic transformations using homogeneous catalysts with high enantioselectivity are described in the literature.

At the present time, only few of the catalysts described above are commercially available. Because some rather special know how is required to prepare and characterize these heterogeneous materials, their application will probably for some time be restricted to the "specialists" among the synthetic organic chemists. We think that many of these heterogeneous chiral catalysts are in principle feasible for technical applications in the pharmaceutical and agrochemical industry. It is in this area that the advantages of heterogeneous catalysts concerning separation and handling properties will be most significant. Therefore, it is here where we expect most progress in the near future. Particularly, we see opportunities in the following directions: immobilization of the exceptionally versatile BINAP ligand [111]; new modifiers for metallic and oxidic catalysts; the enantioselective hydrogenation of aromatic rings. An advantage for the synthetic chemist would be the commercial availability of easy to handle immobilized chiral ligands or catalysts.

ACKNOWLEDGMENTS

We would like to thank our colleagues Martin Studer for valuable discussions and a critical reading of the manuscript and Manfred Müller for support with the art work.

REFERENCES

[1]   For a discussion of the synthetic problems see D. Seebach, Angew. Chem. 102 (1990) 1363.
[2]   For a discussion of the industrial problems see R.A. Sheldon, Chirotechnology, Marcel Decker, Inc., New York, 1993.
[3]   E. Kokufuta, Prog. Polym. Sci., 17 (1992) 647.
[4]   a) H.U. Blaser and M. Müller, Stud. Surf. Sci. Catal. 59 (1991) 73.
      b) H.U. Blaser, Tetrahedron: Asymmetry 2 (1991) 843.
[5]   G. Webb and P.B. Wells, Catal. Today 12 (1992) 319.
[6]   A. Tai and T. Harada, in Y. Iwasawa (Ed.), Taylored Metal Catalysts, D. Reidel, Dordrecht, 1986, p. 265.
[7]   A. Tai, M. Imaida, T. Oda and H. Watanabe, Chem. Lett. (1978) 61.
[8]   A. Tai, N. Morimoto, M. Yoshikawa, K. Uehara, T. Sugimura and T. Kikukawa, Agric. Biol. Chem. 54 (1990) 1753.
[9]   M. Nakahata, M. Imaida, H. Ozaki, T. Harada and A. Tai, Bull. Chem. Soc. Jpn. 55 (1982) 2186.
[10] J. Bakos, I. Toth and L. Marko, J. Org. Chem. 46 (1981) 5427.
[11] Catalogue of Wako Pure Chemicals Industries (Osaka), 22. Edition, p. 471 and 547 (cited in ref. [6]).
[12] E. Broger, Hoffmann-LaRoche, Basel, personal communication.
[13] E.I. Klabunovskii, Kinet. Katal. 33 (1992) 292, English translation p. 233.
[14] H. Brunner, K. Amberger, T. Wischert and J. Wiehl, Bull. Soc. Chim. Belg. 100 (1991) 571 and 585.
[15] M.A. Keane and G. Webb, J. Catal. 136 (1992) 1.
[16] M.A. Keane and G. Webb, J. Mol. Catal. 73 (1992) 91.
[17] M.A. Keane, Catal. Lett. 19 (1993) 197.

[18] A. Tai, T. Kikukawa, T. Sugimura, Y. Inoue, S. Abe, T. Osawa and T. Harada, in "New Frontiers in Catalysis. Proceedings, 10th International Congress on Catalysis, Budapest, 1992", L. Guczi, F. Solymosi and P. Tetenyi, Eds., Akademiai Kiado, Budapest, 1993, p. 2443 and J. Chem. Soc., Chem. Commun. (1991), 795.

[19] H.U. Blaser, H.P. Jalett, D.M. Monti, A. Baiker and J.T. Wehrli, Stud. Surf. Sci. Catal. 67 (1991) 147.

[20] G.H. Sedelmeier, H.U. Blaser and H.P. Jalett; 1986; EP 206'993, assigned to Ciba-Geigy AG.

[21] H.U. Blaser, S.K. Boyer and U. Pittelkow, Tetrahedron: Asymmetry 2 (1991) 721.

[22] G. Bond, P.A. Meheux, A. Ibbotson and P.B. Wells, Catal. Today 10 (1991) 371.

[23] G. Bond, K.E. Simons, A. Ibbotson, P.B. Wells and D.A. Whan, Catal. Today 12 (1992) 421.

[24] K.E. Simons, A. Ibbotson and P.B. Wells, Spec. Publ. - R. Soc. Chem. 114 (Catalysis and Surface Characterization), p. 174 (1992).

[25] H.U. Blaser, M. Garland and H.P. Jalett, to be published in J. Catal. 144 (1993).

[26] R.L. Augustine, S.K. Tanielyan and L.K. Doyle, Tetrahedron: Asymmetry 4 (1993) 1803.

[27] J.L. Margitfalvi, B. Minder, E. Talas, L. Botz and A. Baiker, in "New Frontiers in Catalysis. Proceedings, 10th International Congress on Catalysis, Budapest, 1992", L. Guczi, F. Solymosi and P. Tetenyi, Eds., Akademiai Kiado, Budapest, 1993, p. 2471.

[28] E. Talas, L. Botz, J.L. Margitfalvi, O. Sticher and A. Baiker, J. Planar Chromatogr. 5 .(1992) 28.

[29] K.E. Simons, P.A. Meheux, A. Ibbotson and P.B. Wells, in "New Frontiers in Catalysis. Proceedings, 10th International Congress on Catalysis, Budapest, 1992", L. Guczi, F. Solymosi and P. Tetenyi, Eds., Akademiai Kiado, Budapest, 1993, p. 2317.

[30] O. Schwalm, B. Minder, J. Weber and A. Baiker, in "Abstracts, 2nd G.M. Schwab Symposium, Berlin, 1993". Poster 2.7.

[31] J.L. Margitfalvi, B. Minder, A. Baiker, P. Skrabal and E. Talas, in "Abstracts, Europacat-1, Montpellier, 1993", p. 406.

[32] H.U. Blaser and M. Müller, in "Abstracts, Europacat-1, Montpellier, 1993", p. 408.

[33] S. Bhaduri, V.S. Darshane, K. Sharma and D. Mukesh, J. Chem. Soc., Chem. Commun. (1992) 1738.

[34] H.U. Blaser, H.P. Jalett, D.M. Monti, J.F. Reber and J.T. Wehrli, Stud. Surf. Sci. Catal. 41 (1988) 153.

[35] W.A.H. Vermeer, A. Fulford, P. Johnston and P.B. Wells, J. Chem. Soc., Chem. Commun. (1993) 1053.

[36] H.U. Blaser and H.P. Jalett, Stud. Surf. Sci. Catal. 78 (1993) 139.

[37] P. Gallezot, private communication and K. Nasar, P. Drognat Landre, F. Fache, M. Besson, D. Richard, P. Gallezot and M. Lemaire, in "Abstracts, Europacat-1, Montpellier, 1993", p. 390.

[38] H. Yamashita, Bull. Chem. Soc. Jpn., 61 (1988) 1213.

[39] B.M. Choudary, V.L.K. Valli and A. Durga Prasad, J. Chem. Soc., Chem. Commun. (1990) 1186.

[40] B.M. Choudary, S. Shobha Rani and N. Narender, Catal. Lett. 19 (1993) 299.

[41] R.M. Dessau, US 4'554'262 (11. Nov. 1985) and US 4'666'874 (19. May 1987), assigned to Mobil Oil Corp..

[42] M.E. Davis, Acc. Chem. Res. 26 (1993) 111.

[43] B.M. Choudary and S. Shobha Rani, J. Mol. Catal. 75 (1992) L7.

[44] J.N. Armor, Appl. Catal.78 (1991) 141.

[45] K. Morihara, S. Kawasaki, M. Kofuji and T. Shimada, Bull. Chem. Soc. Jpn. 66 (1993) 906, and references cited therein.

[46] T. Moriguchi, Y. G. Guo, S. Yamamoto, Y. Matsubara, M. Yoshihara and T. Maeshima, Chem. Express 7 (1992) 625.

[47] C. Cativiela, F. Figueras, J.M. Fraile, J.I. Garcia, J.A. Mayoral, E. Pires and A.J. Royo, in "Abstracts, Europacat-1, Montpellier, 1993", p. 395.

[48] For a review see M. Aglietto, E. Chiellini, S. D'Antone, G. Ruggeri and R. Solaro, Pure & Appl. Chem. 60 (1988) 415.

[49] S. Itsuno, M. Sakakura and K. Ito, J. Org. Chem. 55 (1990) 6047.

[50] J. Boulahia, F. Carriere and H. Sekiguchi, Makromol. Chem. 192 (1991) 2969.

[51] J.A.N. Augustyn, B.C.B. Bezuidenhoudt and D. Ferreira, Tetrahedron 46 (1990) 2651.

[52] J.R. Flisak, P.G. Gassman, I. Lantos and W.L. Mendelson, EP 403'252 (19. Dec. 1990), assigned to SmithKlien Beecham Corp.. for the structure of the active compound see J.G. Gleason et al., J. Med. Chem. 30 (1987) 959.

[53] A. Hanabusa, J. Ichihara, H. Tanaka and T. Kimura, JP O4 29'965 (31. Jan. 1992), assigned to Katakura Chikkarin Co., Ltd.. For an abstract see CA 117 (1992) 48120a.

[54] T. Sugimoto, T. Kokubo, J. Miyazaki, S. Tanimoto and M. Okano, Bioorg. Chem. 10 (1981) 311.

[55] S. Colonna, S. Banfi, F. Fontana and M. Sommaruga, J. Org. Chem. 50 (1985) 769.

[56] T. Sugimoto, T. Kokubo, Y. Matsumura, J. Miyazaki, S. Tanimoto and M. Okano, Bioorg. Chem. 10
      (1981) 104.
[57] R. Annunziata, S. Banfi and S. Colonna, Tetrahedron Lett. 26 (1985) 2471.
[58] S. Colonna and A. Manfredi, Tetrahedron Lett. 27 (1986) 387.
[59] M.E. Wilson and G.M. Whitesides, J. Am. Chem. Soc. 100 (1978) 306.
[60] H. Alper and N. Hamel, J. Chem. Soc., Chem. Commun. (1990) 135.
[61] T. Kokubo, T. Sugimoto, T. Uchida, S. Tanimoto and M. Okano, J. Chem. Soc., Chem. Commun.
      (1983) 769.
[62] B. Pispisa, G. Paradossi, A. Palleschi and A. Desideri, J. Phys. Chem. 92 (1988) 3422; B. Pispisa, A.
      Palleschi and G. Paradossi, J. Mol. Catal. 42 (1987) 269; and references cited therein.
[63] L. Verbit, T. R. Halbert and R.B. Patterson, J. Org. Chem. 40 (1975) 1649.
[64] W.H. Pirkle and P.L. Rinaldi, J. Am. Chem. Soc. 99 (1977) 3510.
[65] V.A. Pavlov, N.I. Spitsyna and E.I. Klabunovskii, Izv. Akad. Nauk SSSR, Ser. Khim. (1983) 1653,
      English translation p. 1501.
[66] H. Danda, Synlett (1991) 263.
[67] Y. Becker, A. Elgavi and Y. Shvo, GB 2'227'429 (5. Dec. 1988), assigned to Bromine Compounds
      LTd..
[68] R. Selke, K. Häupke and H.W. Krause, J. Mol. Catal. 56 (1989) 315.
[69] F.R. Hartley, Supported metal complex catalysts, in: F.R. Hartley (Ed.), The Chemistry of Metal-
      Carbon Bond, John Wiley & Sons Ltd, Vol. 4, p. 1163, (1987).
[70] J. Hetflejs, Stud. Surf. Sci. Catal. 27 (1986) 497-515.
[71] J.K. Stille., Organic Synthesis via Polymer-supported Transition Metal-Catalysts. Reactive Polymers
      10 (1989) 165.
[72] T. Matsuda and J.K. Stille, J. Am. Chem. Soc. 100 (1978) 268.
[73] G.L. Baker, S.J. Fritschel and J.K. Stille, J. Org. Chem. 46 (1981) 2960.
[74] R. Deschenaux and J.K. Stille, J. Org. Chem. 50 (1985) 2299.
[75] A. Corma, M. Iglesias, C. del Pino and F. Sanchez, J. Chem. Soc., Chem. Commun. (1991) 1253.
[76] A. Corma, M. Iglesias, C. del Pino and F. Sanchez, J. Organomet. Chem. 431 (1992) 233.
[77] A. Corma, M. Iglesias, M.V. Martin, J. Rubio and F. Sanchez, Tetrahedron: Asymmetry 3 (1992) 845.
[78] B. Pugin and M. Müller, in: M. Guisnet et al. (Ed.), Heterogeneous Catalysis and Fine Chemicals III,
      Stud. Surf. Sci. and Cat., 78 (1993) 107.
[79] B.L. Booth, G.C. Ofunne, C. Stacey and P.J.T. Tait, J. Organomet. Chem., 315 (1986) 142.
[80] W. Dumont, J.C. Poulin, P.T. Dang and H.B. Kagan, J. Am. Chem. Soc. 95 (1973) 8295.
[81] J.M.J. Frechet, G.D. Darling, S. Itsuno, P.-Z. Lu, V. de Meftahi and W.A. Rolls, Pure & Appl. Chem.,
      60 (1988) 353.
[82] J. Lieto, D. Milstein, R.L. Albright, J.V. Minkiewicz and B.C. Gates, Chemtech (1983) 46.
[83] H. Engelhardt and P. Orth, J. Liqu. Chromatogr., 10 (1987) 1999.
[84] M. Cerny, Coll. Czechoslov. Chem. Commun., 42 (1977) 3069.
[85] U. Nagel and E. Kinzel, J. Chem. Soc., Chem. Commun. (1986) 1098.
[86] S. Itsuno, S. Yoshiki, I. Kiochi, M. Toshihiro, N. Seiichi and J.M.J. Frechet, J. Org. Chem. 55 (1990)
      304.
[87] T. Hayashi, Tetrahedron Lett. 21 (1980) 80.
[88] K.J. Stille, J. Macromol. Sci.-Chem. A21 (1984) 1689.
[89] U. Nagel, H. Menzel, P.W. Lednor, W. Beck, A. Guyot and M. Bartholin, Z. Naturforsch. B: Anorg.
      Chem., Org. Chem., 36B (1981) 578.
[90] K. Achiwa, Heterocycles 9 (1978) 1539.
[91] K. Achiwa, Chem. Lett. 8 (1978) 905.
[92] G.L. Baker, S.J. Fritschel and J.R. Stille, J. Org. Chem. 46 (1981) 2954.
[93] K.B. Moon and K.B. Sharpless, Tetrahedron Lett. 31 (1990) 3003.
[94] R. Selke and M. Capka, J. Mol. Catal. 63 (1990) 319.
[95] I. Toth and B.E. Hanson, J. Mol. Catal. 71 (1992) 365.
[96] N. Ishizuka, M Togashi, M. Inoue and S. Enomoto, Chem. Pharm. Bull. 35 (1987) 1686.
[97] H. Brunner, E. Bielmeier and J. Wiehl, J. Organomet. Chem. 384 (1990) 223.
[98] M. Inoue, K. Ohta, N. Ishizuka and S. Enomoto, Chem. Pharm. Bull. 31 (1983) 3371.
[99] M. Mazzei, W. Marconi and M. Riocci, J. Mol. Catal. 9 (1980) 381.
[100] J. Kaschig, European Patent 0,252,994,A1, 1988; Chem. Abstr., (1988) 109, 55454a.   see also: G.
      Zassinovich, G. Mestroni and S. Gladiali, Chem. Rev. 92 (1992) 1051.
[101] G. Parrinello and J.K. Stille, J. Am. Chem. Soc. 109 (1987) 7122.
[102] J.K. Stille, H. Su, G. Parrinello and L.S. Hegedus, Organometallics 10 (1991) 1183.

[103] Y. Gao, R.M. Hanson, J.M. Klunder, S.Y. Ko, H. Masamune and K.B. Sharpless, J. Am. Chem. Soc. 109 (1987) 5765.

[104] K.B. Sharpless, W. Amberg, Y.L. Bennani, G.A. Crispino, J. Hartung, K.-S. Jeong, H.-L. Kwong, K. Morikawa, Z.-M. Wang, D. Xu and X.L. Zhang, J. Org. Chem. 57 (1992) 2768.

[105] S. Itsuno, Y. Sakurai, K. Ito and A. Hirao, Polymer 28 (1987) 1005.

[106] S. Itsuno, K. Ito, T. Maruyama, N. Kanda, A. Hirao and S. Nakahama, Bull. Chem. Soc. Jpn. 59 (1986) 3329.

[107] S. Itsuno and J.M.J. Frechet, J. Org. Chem. 52 (1987) 4140.

[108] G. Parrinello, R. Deschenaux and J.K. Stille, J. Org. Chem. 51 (1986) 4189.

[109] S.A. Matlin, W.J. Lough, L. Chan, D.M.H. Abram and Z. Zhou, J. Chem. Soc., Chem. Commun. (1984) 1038.

[110] M.P. Doyle, M.Y. Eismont, D.E. Bergbreiter and H.N. Gray, J. Org. Chem. 57 (1992) 6103.

[111] C. Rosini, L. Franzini, A. Raffaelli and P. Salvadori, Synthesis (1992) 503.

SECTION II

HYDROGENATION SYSTEMS :

TOWARD A BETTER UNDERSTANDING

MODIFICATION OF SUPPORTED AND UNSUPPORTED NICKEL CATALYSTS BY α–AMINO AND α–HYDROXY ACIDS FOR CHIRAL REACTIONS

Geoffrey Webb

Department of Chemistry
The University
Glasgow G12 8QQ, Scotland

INTRODUCTION

The use of heterogeneous metal catalysts to effect asymmetric reactions is not new, dating back to work in the 1930's by Schwab [1,2] and by Lipkin and Stewart [3]. However, it is only in the past 20 years that the subject has been studied in any detail.

The requirement of asymmetric activity is the existence of a chiral environment on the catalyst surface and this may be created in one of two ways. First, by supporting the metal on a chiral support and second, by adsorption of a chiral modifier on the active phase of a conventional metallic catalyst. Whilst examples of both approaches are extant, the latter approach is the one which has been more commonly used.

Whilst as detailed in the recent reviews by Blaser [4] and Blaser and Muller [5], many reactions have been identified, only two systems have been subjected to detailed study, both of which have used the approach of modifying the catalyst by the adsorption of a chiral modifier. These systems are (a) modification of nickel catalysts, usually Raney Ni, by α–hydroxy or α–amino acids and (b) modification of platinum catalysts with the cinchona alkaloid derivatives.

From these studies one of the most important features to emerge is that such reactions are not only very dependent on the modifier used, but are also very reactant specific. Thus, modified Ni catalysts are specific for the hydrogenation of β–keto esters, of which methyl acetoacetate has been most commonly used;

$$H_3COCH_2COOCH_3 \quad \rightarrow \quad CH_3C^*H(OH)CH_2COOCH_3$$

with (R)–hydroxy or (S)–amino acids yielding (R)–product and (S)–hydroxy or (R)–amino acids yielding (S)–product.

On the other hand cinchona–modified Pt catalysts are specific for the hydrogenation of α–ketoesters, for which methyl pyruvate is the usual model reactant.

$$CH_3COCOOCH_3 \quad \rightarrow \quad CH_3C^*H(OH)COOCH_3$$

with chinchonidine or quinine yielding (R)–methyl lactate and cinchonine or quinidine yielding (S)–methyl lactate.

In our studies which are described in this review, we have concentrated on the modification of nickel catalysts by α–hydroxy acids (with particular reference to tartaric acid) and α–amino acids (in particular alanine) and have used the hydrogenation of methylacetoacetate as our model enantioselective reaction.

At the time of commencement of our studies, a review of the literature revealed that whilst many different facets of the reaction had been investigated, particularly for Raney Ni by Izumi and co–workers [6] and for supported Ni catalysts by Sachtler et al. [7–12] and by Nitta et al. [13–16], there was no consensus view as to the source of enantiodifferentiation on the modified catalyst surfaces, although there was a general agreement that this depended critically on the method of catalyst preparation and the nature of the modification process. Among the various factors which have been claimed to be of importance were : (a) Metal dispersion was important – larger metal particles being better than small ones [16,17]– whilst with supported Ni catalysts higher metal loadings improve enantioselectivity [5]; (b) the preferred structure for the modifier is RCHXCOOH (X = NH$_2$ or OH), molecules containing two chiral centres being better than those with one. Increasing bulkiness of R was favourable if X=NH$_2$, but unfavourable when X= OH : (c) both the temperature and pH of modification were important, although different studies claim different effects; (d) addition of NaX (X= Cl or Br) may be advantageous to increasing enantioselectivity, and (e) some nickel leaching occurs during the modification process, although no account appears to have been taken of the effects of this on the enantiodifferentiating abilities of the catalyst.

As to the mechanism of enantiodifferentiation, the main points of debate still remain as to (a) the nature of the surface complex produced by the modification with, for example tartaric acid modification, ideas ranging from the formation of an adsorbed species capable of forming a complex with the reactant molecule [5], to the formation of Na–Ni–tartrate Ni complexes on the catalyst, through to the formation of heteroligand chelate complexes [18,19] of the type NiL$_1$L$_2$ (where L$_1$= substrate and L$_2$= ditartrate ion), (b) the effects of nickel leaching during modification, but only in the context of the possible role of complexes resulting from this leaching, (c) the role, if any, of nickel complexes leached from the catalysts during modification. Thus, it has been claimed that Ni(alanine)$_2$ has a similar activity/enantioselectivity to alanine modified Raney nickel [20], and (d) the role of sodium halide promoters.

Against this background we have carried out a comprehensive study of the modification of both Raney Ni and supported Ni catalysts with the objective of (a) understanding in detail the modification process and its effects on the catalyst, and (b) obtaining a detailed understanding of the effects of the modification process on catalytic activity and enantioselectivity using the hydrogenation of methylacetoacetate (MAA) to 3–methyl hydroxybutyrate (MHB) as a model reaction system.

Our studies have been carried out using both alanine and tartaric acid as modifiers. Most of the examples quoted below are with reference to tartaric acid since this is the one which has been most widely used in previous studies. However, the general effects are similar for both hydroxy– and amino–acid modified systems.

EXPERIMENTAL DETAILS

Catalysts

Raney Ni catalysts were prepared by standard methods involving the digestion of Raney Ni/Al alloy in aqueous NaOH solution, followed by washing in water, followed by methanol and then either toluene or n–butanol, depending on which was used as solvent for the hydrogenation reactions.

Nickel/silica catalysts with Ni loadings of between 1.5 and 24.3% w/w Ni were prepared using aqueous nickel nitrate and the non–porous Cab–O–Sil 5M silica as support by a homogeneous precipitation/deposition technique developed by van Dillen *et al.* [21]. The supported catalysts were characterised by temperature programmed reduction, thermogravimetric analysis, high resolution electron microscopy and carbon monoxide chemisorption. Some experiments were also carried out using Ni/γ–alumina catalysts, although it quickly became apparent that, at comparable metal loadings, these catalysts were very inferior to the Ni/silica catalysts, both in terms of hydrogenation activity and enantioselectivity for MAA hydrogenation.

Reactions

With both Raney Ni and Ni/silica catalysts, hydrogenation of MAA was carried out in a standard liquid phase hydrogenation system at 1 atm. pressure and either toluene (Raney Ni) or n–butanol (Raney Ni and Ni–silica) as solvent. To avoid problems associated with the diffusion of dihydrogen through the reaction medium, the dihydrogen was continuously bubbled through the catalyst/liquid reactant suspension, usually at a rate of 60 cm^3 min^{-1}. Some reactions were also carried out under static conditions at elevated hydrogen pressures ($\leq$90 Bar) using Raney Ni. Overall product yields were determined by HPLC, using a Pirkle type 1A 5μ reversible column with a 10% IPA : 90% hexane mixture as the mobile phase. The optical yields were determined by polarimetry using the sodium D–line.

$$\text{Optical Yield (O.Y)} = \frac{[\alpha]_D^T}{[\alpha]_0^T}$$

where $[\alpha]_D^T$ = specific rotation of product solution at T K;

$[\alpha]_0^T$ = specific rotation of pure enantiomer at T K;

and

$$\text{Enantiomeric excess (e.e.)\%} = 100.\frac{([(R)-(-)-MHB]-[(S)-(+)-MHB])}{([(R)-(-)-MHB]+[(S)-(+)-MHB])}$$

Modification Process

Preliminary investigations were carried out to establish a modification process which would lead to reproducible results and from these the following procedure was adopted throughout the studies.

1. The freshly reduced catalyst was contacted, under N_2, with the aqueous modifier solution (typically 100 cm^3) at ambient temperature.
2. The pH of the modifier solution was then adjusted to the required value (between 1.5 and 11.8) using aqueous NaOH.

3. The solutions contacting the catalyst were further purged with N_2 to remove entrapped air bubbles. (This was found to be essential).
4. The temperature of the modifier solution/catalyst suspension was adjusted to the required value (between 273 and 373 K).
5. The modification was carried out for the required time under constant agitation (600 r.p.m.). This latter step being carried out in either a nitrogen or air atmosphere.

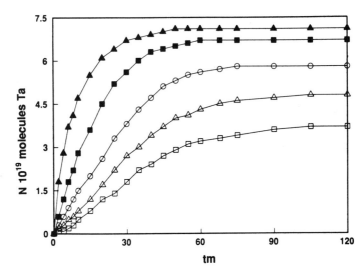

Figure 1. Amount of tartaric acid adsorbed on a 15.2% Ni/silica catalyst as a function of time of modification (t_{mod}) at 273 K (□), 298 K (Δ), 323 K (O), 343 K (■) and 373 K (▲) ($[TA]_{initial} = 3.3 \times 10^{-2}$ mol dm^{-3}; pH = 5.1)

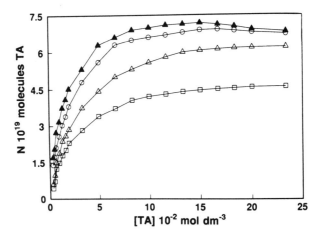

Figure 2. Amount of tartaric acid adsorbed on a 11.9% Ni/silica catalyst as a function of $[TA]_{initial}$ at 273 K (□), 306 K,(Δ), 343 K (O) and 373 K (▲) (t_{mod}= 2 hr.; pH = 5.1).

An HPLC technique was also developed to permit the determination of the extent of tartaric acid (TA) adsorption during the modification process. Typical results for the amounts of tartaric acid adsorbed at various times in the modification process and at various temperatures using a constant $[TA]_{initial}$ of 3.3 x 10^{-2} mol dm^{-3} and a constant initial modifier pH = 5.1 are shown in figure 1, whilst figure 2 shows the variation in the amount of tartaric acid adsorbed as a function of the $[TA]_{initial}$ in the modifier solution at similar temperatures using a constant pH$_{initial}$ = 5.1 and a modification time of 2 hr.

From these it can be seen that the amount of TA adsorbed is dependent on the time of the modification and, in all cases, the maximum surface coverage of TA for a given $[TA]_{initial}$ requires periods in excess of 1 hr.

During the modification process, extensive nickel leaching was observed to occur. Measurement of the amounts of nickel leached from the catalyst during modification are shown in figures 3 and 4, from which it is clear that, not surprisingly, the amount leached is dependent on the $[TA]_{initial}$, temperature of modification ($T_{mod.}$) and the time of modification ($t_{mod.}$).

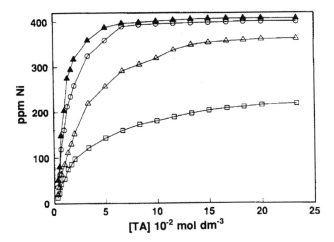

Figure 3. Effect of $[TA]_{initial}$ on the amount of nickel leached from a 11.9% Ni/silica catalyst during modification at 273 K (□), 306 K (Δ), 343 K (O) and 373 K (▲). ($t_{mod.}$= 2 hr.; pH$_{initial}$ = 5.1).

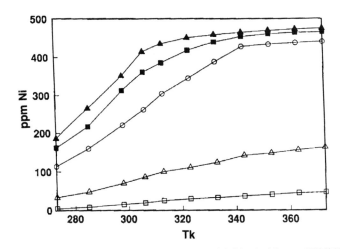

Figure 4. Effect of temperature of modification on the amount of nickel leached from a 15.2% Ni/silica catalyst at various $[TA]_{initial}$(10^{-2} mol dm^{-3}); 0.3 (□), 0.7 (Δ), 3.3 (O), 6.7 (■) and 13.3 (▲). (t_{mod} = 2 hr., pH$_{initial}$ = 5.1).

Characterisation of the catalysts in the freshly reduced and the modified states, after treatment at elevated temperature (583 K) in flowing helium to remove any desorbable modifier, was carried out by HRTEM to determine nickel particle size distributions, as shown in figure 5, and by carbon monoxide chemisorption at 273 K, a method which has been shown to give comparable results to those obtained by dioxygen chemisorption [22], to determine metal dispersions (table 1). The results clearly show that during the modification process the small nickel particles are preferentially leached from the catalyst and, as a consequence, the extent of leaching was less with the catalysts with the higher nickel loading, since these also have the lowest dispersion in the freshly reduced state.

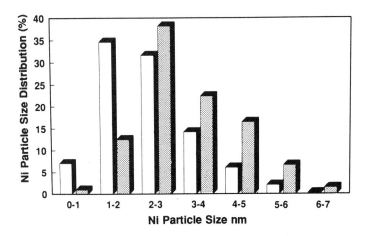

Figure 5. Nickel particle size distributions in a 11.9% Ni/silica catalyst before (open bar) and after modification (hatched bar). (For modification $[TA]_{initial}= 6.7 \times 10^{-3}$ mol dm^{-3}; $T_{mod}= 343$ K; $t_{mod}= 2$ hr. and $pH_{initial}= 5.1$)

Table 1. Effects of modification on Ni/Silica catalysts.

% w/w Ni	%age Ni Leached	D^a Before Modification	D^a After Modification
1.5	51	73	39
6.1	29	54	36
11.9	19	40	33
15.2	16	33	28
20.3	14	27	25
24.7	13	24	23

[a] Dispersions determined from CO chemisorption measurements, calculated on the basis of $(Ni_{surface}/CO_{ads.}) = (2/1)^{23}$

The amount of TA adsorbed by the catalyst is also dependent on the initial pH of the modifier solution. As shown in figure 6, increasing the pH increases the extent of TA adsorption up a pH of approximately 5. Beyond this value, where the TA or alanine is present predominantly as the mono sodium salt, the amounts adsorbed decrease with increasing pH, implying a weaker interaction between the modifier and the nickel surface under these conditions.

From figure 6 it can be seen that the maximum amount of TA adsorbed occurs at an initial modifier solution pH corresponding to the isoelectric point (pH = 5.1) of tartaric acid solution. Similarly with alanine as the modifier, a maximum in the amount adsorbed was observed at its isoelectric point (pH = 6.0) [24]. With both tartaric acid and alanine modifiers, the pH of the solution was observed to increase during the modification, particular at $pH_{initial} < 7$. However, following the method devised by Wittmann *et al.* [25] for Raney nickel, whereby the pH was kept constant throughout the modification and which these authors claimed resulted in a marked improvement in enantioselectivity, had no significant effect on either the activity or enantioselectivity of our silica–supported nickel catalysts.

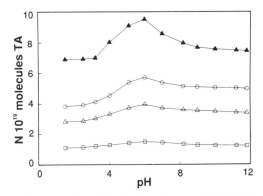

Figure 6. Effect of $pH_{initial}$ on the extent of tartaric acid adsorption on a 15.2% Ni/silica catalyst at various $[TA]_{initial}$. (($\square$) T_{mod}=273 K, $[TA]_{initial}$= 6.7 x 10^{-3} mol dm^{-3}; (Δ) T_{mod}= 273 K, $[TA]_{initial}$= 3.3 x 10^{-2}mol dm^{-3}; (O) T_{mod}= 273 K, $[TA]_{initial}$= 6.7 x 10^{-2} mol dm^{-3}; ($\blacktriangle$) T_{mod}= 373 K, $[TA]_{initial}$= 6.7 x 10^{-2} mol dm^{-3}).

Direct comparison of the amounts of TA adsorbed with the amounts of CO adsorbed on both the freshly reduced and modified catalysts leads to the conclusion that the TA (and alanine) is adsorbed on the metal, rather than on the metal and support.

Hydrogenation Reactions of Methylacetoacetate (MAA).

As noted above, in general, reactions were carried out at 1 atm pressure, and figure 7 shows typical time vs. product yield curves for the reaction on an unmodified and a modified 15.2% Ni/silica catalyst at 343 K.

Figure 8 shows the variation in the reaction rate as a function of the initial MAA substrate/nickel ratio for an unmodified catalyst (lower curve) and for the same catalyst modified with tartaric acid (upper curve), the results for the latter catalyst being corrected for the amount of nickel leached from the catalyst during the modification process.

From figures 7 and 8 it is particularly notable that the modified catalyst, after correction for the amount of Ni removed in the leaching process, is of *significantly higher activity* than the unmodified catalyst. This appears to be a typical phenomenon for chiral modified catalysts since similar behaviour is observed with alanine modified nickel catalysts [24] and Wells *et al.* [26] have reported quite spectacular increases in reaction rates for methyl pyruvate hydrogenation with cinchonidine modified platinum catalysts. In contrast with the activity, the enantioselectivity appears to be almost independent of the substrate/catalyst ratio in the reaction as shown in figure 9.

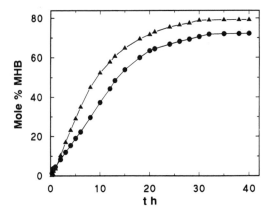

Figure 7. Variation in product yield with time for the hydrogenation of MAA over modified (($\blacktriangle$), $(MAA/Ni)_{initial}$= 57) and unmodified (($\bullet$), $(MAA/Ni)_{initial}$= 46) 11.9% w/w Ni/silica catalyst at 343 K. (Modification : $[TA]_{initial}$= 6.7×10^{-3} mol dm^{-3}; T_{mod}= 343 K; $pH_{initial}$= 5.1 and t_{mod}= 2 hr.)

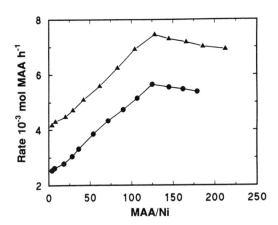

Figure 8. Variation in the rate of reaction as a function of the MAA reactant/catalyst ratio on a 15.2% Ni/silica catalyst, using a constant initial weight of catalyst. (($\blacktriangle$) modified $[TA]_{initial}$= 6.7×10^{-3}, $pH_{initial}$= 5.1, t_{mod}= 2 hr., T_{mod}= 343 K; ($\bullet$) unmodified).

The enantioselectivity observed with the modified catalysts also depends on the reaction temperature, passing through a maximum at *ca.* 340–360 K dependent on the metal loading of the catalyst , as shown in figure 10 .

Determination of the amounts of MAA adsorbed on variously modified catalysts, as shown in figure 11, shows that (a) the total amount of MAA adsorbed under reaction conditions is always in considerable excess of what can be accommodated on the metal component itself, implying adsorption on the support; (b) as with the activity the amount of MAA adsorbed passes through a maximum in the temperature range 320–350 K; (c) adsorbed TA effectively reduces the amount of MAA adsorbed.

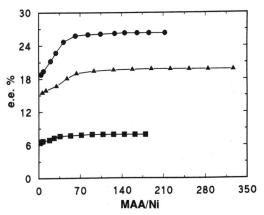

Figure 9. Variation of the enantiomeric excess (e.e.%) with MAA/nickel ratio using a modified 15.2% Ni/silica catalyst ([TA]$_{initial}$ = 6.7 x 10^{-2} mol dm^{-3} (▲); 6.7 x 10^{-3} mol dm^{-3} (●) or 3.3 x 10^{-3} mol dm^{-3} (■) at pH$_{initial}$= 5.1, t$_{mod}$= 2 hr. and T$_{mod.}$= 343 K).

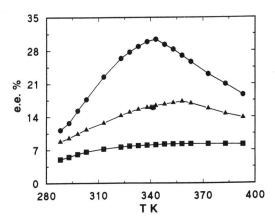

Figure 10. Variation in the enantiomeric excess (e.e.%) with reaction temperature for modified 6.1% w/w Ni/silica catalyst ((●), [TA]$_{initial}$= 2x10^{-2} mol dm^{-3}, (MAA/Ni)$_{initial}$= 81); 11.9% w/w Ni/silica ((▲), [TA]$_{initial}$= 2x10^{-2}mol dm^{-3}, (MAA/Ni)$_{initial}$= 67) and 24.3% w/w Ni/silica (EURONI–1) ((■), [TA]$_{initial}$= 1x10^{-2} mol dm^{-3} (MAA/Ni)$_{initial}$= 27). For each catalyst, t$_{mod}$= 2 hr. and pH$_{initial}$ = 5.1)

Attempts have also been made to determine the optimum conditions for the modification process to obtain maximum enantioselectivity and maximum activity. Figure 12 shows that with both tartaric acid and alanine modification, the enantiomeric excess passes through a maximum with increasing initial concentration of modifier. This maximum occurs at initial modifier concentrations of *ca.* 0.013 mol dm^{-3} for TA and *ca.* 0.11 mol dm^{-3} for alanine. Combination of these results with the CO chemisorption data, data on the extent of nickel leaching and the HPLC data on the numbers of TA and alanine molecules adsorbed, allows the calculation of the surface coverages corresponding to maximum enantioselectivity. These are 0.2 and 0.7 for TA and alanine respectively. For TA, this value appears to be almost independent of the temperature of modification. These results also show that, contrary to claims made for Raney Ni catalysts [27], alanine modification can lead to increased enantioselectivity as compared with tartaric acid modification.

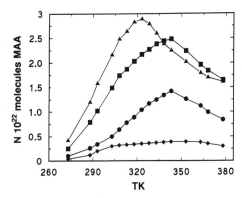

Figure 11. Variation in the number of molecules MAA adsorbed on unmodified (▲) and variously modified 20.3% w/w Ni/silica catalyst samples at various temperatures. (All modifications carried out with t_{mod}= 2 hr., $pH_{initial}$= 5.1. (■) $[TA]_{initial}$= 6.7 x 10^{-3} mol dm^{-3}, T_{mod}= 343 K, $(MAA/Ni)_{initial}$= 31; (●) $[TA]_{initial}$= 5.0 x 10^{-2} mol dm^{-3}, T_{mod}= 343 K, $(MAA/Ni)_{initial}$= 40; (♦) $[TA]_{initial}$= 8.3 x 10^{-2} mol dm^{-3}, T_{mod}= 373 K, $(MAA/Ni)_{initial}$= 64).

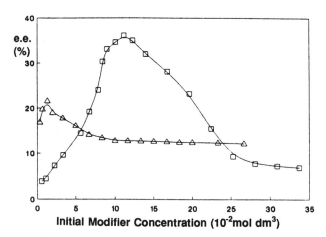

Figure 12. Variation of the enantiomeric excess (e.e.%) with initial concentration of TA modifier (△) and alanine modifier (□) for the hydrogenation of MAA over a modified 11.9% w/w Ni/silica catalyst ($T_{reaction}$= 343 K; $t_{reaction}$= 32 hr.; T_{mod} = 343 K; t_{mod} = 2 hr.; $pH_{initial}$= 5.1)

The use of NaBr as a co–modifier has also been investigated as part of the optimisation process. As shown in table 2, for TA modified nickel/silica catalysts, addition of NaBr to the modifier solution resulted in an increase in the enantiomeric excess, but a decrease in extent of MAA conversion to MHB, although the effects are not as marked as those observed with Raney nickel catalysts, both in previous work [5,28,29] and in our own studies (table 3), where the effect of the Br⁻ ion may be attributed to a poisoning of any residual aluminium which is known to have a detrimental effect on the enantioselectivity.

The promoting effect of addition of NaBr can be attributed to the combined effects of a poisoning of bare nickel sites on the modified catalyst and a beneficial regulation of the adsorbed modifier concentration. Similar poisoning of bare nickel sites on the modified catalyst can be effected by addition of thiophene, which is well established as a poison for nickel hydrogenation catalysts. We have found that such additions can result in an increase in enantiomeric excesses from *ca.* 30% to *ca.* 47%, although the activity is reduced by *ca.* 25% for reactions carried out under comparable conditions.

Table 2. Effect of NaBr co–modifier on extent of TA adsorption, extent of Ni leaching, activity and enantioselectivity.

[TA]	NaBr	$TA_{adsorbed}$	Ni leached	MHB yield	e.e.
10^{-2} mol dm^{-3}	Added	10^{19} molecules.g^{-1} cat	ppm	mol %	%
0.3	Yes	1.2	32	75	24.4
	No	1.4	36	79.4	23.6
0.7	Yes	1.8	106	74.2	32.0
	No	2.1	120	78.9	27.3
3.3	Yes	4.4	296	73.5	14.1
	No	4.8	325	76.2	11.8
6.7	Yes	5.9	352	72.1	11.4
	No	6.3	390	73.9	8.3
13.3	Yes	6.3	356	72.0	11.4
	No	6.8	398	72.8	8.0

Catalyst = 11.9% Ni/Silica
Modification Conditions : T_{mod}= 343 K; t_{mod} = 2 hr; pH$_{initial}$= 5.1, [NaBr] = 2 g in 100 cm^{-3} TA solution.
Reaction Time = 32 hr. at 343 K.
For the unmodified catalyst under similar reaction conditions, MHB (mol %) = 71.8

Table 3. Effect of NaBr on activity and enantioselectivity of Raney Nickel catalysts

H$_2$ Pressure	NaBr added	Mole % MHB	e.e. (%)
Atmospheric	Yes	8.0	50.0
Atmospheric	No	27	32.0
90 atm.	Yes	100	85.0
90 atm.	No	100	91.5

Modification conditions : $T_{mod.}$ = 273 K; [TA] = 6.7 x 10^{-3} mol dm^{-3} ; t_{mod} = 2 hr.
Reaction conditions : Solvent = Toluene; Time of reaction = 20 hr.

Investigations of the effects of the solvent on the activity and enantioselectivity of the modified catalysts revealed the following decreasing order of effectiveness :

n–alkanols > methylpropionate $\cong$ ethyl acetate >> THF>> toluene $\cong$ CH$_3$CN

Clearly, the reaction is favoured by the use of semi–polar solvents and it has been established that a relationship exists between the activity and dielectric constant of the solvent. Although it is not yet clear as the origins of this effect, our observations suggest that the adsorbed species involved in the reaction may have some degree of ionicity.

In view of the various suggestions that, on the modified catalysts, the seat of the chirality may reside in a nickel–modifier complex [10,18,20], separate experiments have been carried our to determine the activity of nickel (II) tartrate, prepared according to the method described by Hoek and Sachtler [10] and of the complex leached during the modification process and subsequently precipitated from the post–modification solution by addition of a 70:30 n–butanol : methanol mixture, as *heterogeneous* catalysts for MAA hydrogenation under comparable conditions to those used for the tartaric acid modified catalysts. The results are summarised in table 4. from which it can be seen that whilst both the nickel(II) tartrate and the leached complex effectively catalyse the reaction and exhibit similar enantioselectivities and activities, these enantioselectivities and activities are substantially less than those observed with a TA–modified Ni/silica catalyst. Clearly, whilst these are not, in themselves, the sole seat of the chirality exhibited by the tartaric acid modified catalysts, equally the presence of bare nickel on a modified catalyst, as has been suggested as being a requirement for activation of hydrogen is not a necessary prerequisite for the asymmetric hydrogenation. In this context it is interesting to note that during the hydrogenation of MAA on both alanine and tartaric acid modified catalysts which, before use, have been subjected to extensive washing to remove all traces of "leached nickel", a very small amount of a nickel species can be observed in solution, which after decanting the nickel catalyst and addition of further MAA possesses a small enantioactivity for further hydrogenation.

Table 4. Enantioselectivities and activities of nickel(II) tartrate and "leached nickel" at various reaction temperatures.

T (K)	Nickel Tartrate		Leached Nickel	
	Mol % MHB	e.e. (%)	Mol % MHB	e.e.(%)
308	20.1	10	12.1	13
323	48.8	14	44.3	19
333	60.2	15	61.5	20
343	79.1	15	81.4	18
353	85.1	12	88.6	14
368	93.4	8	97.3	8

CONCLUSIONS

The results presented above clearly demonstrate the extreme complexity of the chirally modified nickel system, both in terms of the modification process itself and the nature of the interaction between the modified catalyst and the organic reactant. Our studies using the nickel (II) tartrate complex and the complex extracted from the post modification solution show that, whilst both of these complexes possess some activity and enantioselectivity, they are not the seat of the chirality, which is associated with the surface of the modified catalyst.

The modification process results not only in the adsorption of the modifier on the nickel surface, but also in a modification of the nickel metal in terms of an appreciable decrease in the dispersion due to the leaching of the smaller nickel particles during the treatment with the corrosive α–hydroxy or α–amino acid. It is also clear that with both tartaric acid and alanine that there is an optimum surface coverage for maximum enantioselectivity and that under the conditions where this is achieved a considerable fraction of the nickel surface is unaffected and is available for the racemic hydrogenation of the organic substrate. Poisoning of this part of the surface following modification appears to offer a possible route to improving the enantiodifferentiating ability of the modified catalyst. In agreement with results from studies of other chirally modified systems, our results show that the activity of tartaric acid or alanine modified nickel catalysts for methyl acetoacetate hydrogenation is invariably higher than that of the unmodified nickel catalysts.

Collectively, our results can be interpreted in terms of a mechanism in which, from modelling experiments, optimum activity and enantioselectivity is associated with conditions under which a 1:1 interaction between the adsorbed modifier and the organic reactant can occur. High modifier surface concentrations, where the adsorbed reactant can interact with more than one adsorbed modifier molecule result in a decrease in enantioselectivity. It is apparent that, in order to obtain a fuller understanding of the chiral site on the modified surface it is necessary to establish the nature of the adsorbed state of the modifier and of the modifier/adsorbed reactant complex. Unfortunately, to date, our attempts to achieve this, by the use of infrared spectroscopy, these have not proved to be successful.

ACKNOWLEDGEMENTS

The author wishes to thank Dr. M.A. Keane, Dr. S. Maberley and Dr. A. Bennett for the contributions made to these studies. Thanks are also due to the Science and Engineering Research Council (Interfaces and Catalysis Initiative) and the Carnegie Trust for Scotland for their generous financial support for these studies.

REFERENCES

[1] G.M. Schwab and L. Rudolph, *Naturwiss*, **20**, 262 (1932)
[2] G.M. Schwab, F. Rost and L. Rudolph, *Kolloid–Zeitschrift*, **68**, 157 (1934)
[3] D. Lipkin and T.D. Stewart, *J. Amer. Chem. Soc..*, **61**, 3295, 3297 (1939)
[4] H–U. Blaser, *Tetrahedron, Asymmetry*, **2**, 843 (1991).
[5] H–U. Blaser and M. Muller, in "Heterogeneous Catalysis and Fine Chemicals" (M. Guisnet *et al.* Eds.), Elsevier, Amsterdam, 1991) p. 73.
[6] Y. Izumi, *Advan. Catal.*, **32**, 215 (1983)
[7] D.R. Richards, H.H. Kung and W.M. H. Sachtler, *J. Mol. Catal.*, **36**, 329 (1986)
[8] L. Fu, H.H. Kung and W.M. H. Sachtler, *J. Mol. Catal.*, **42**, 29 (1987)
[9] L.J. Bostelaar and W.M.H. Sachtler, *J. Mol. Catal.*, **27**, 387 (1984)
[10] A. Hoek and W.M.H. Sachtler, *J. Catal.*, **58**, 276 (1979)
[11] J.A. Groenewegen and W.M.H. Sachtler, *Proc. 6th. Intern. Congr. Catal., London, 1976*, p. 1014.
[12] A. Hoek, H.M. Woerde and W.M.H. Sachtler, *Proc. 7th. Intern. Congr. Catal., Tokyo, 1980*, p. 376.
[13] Y. Nitta, F. Sekine, J. Sasaki, T. Imanaka and S. Teranishi, *J. Catal.*, **79**, 211 (1983)
[14] Y. Nitta, T. Utsumi, T. Imanaka and S. Teranishi, *J. Catal.*, **101**, 376 (1986)
[15] Y. Nitta, O. Yamenishi, F. Sekine, T. Imanaka and S. Teranishi, *J. Catal.*, **79**, 475 (1983)
[16] Y. Nitta, T. Imanaka and S. Teranishi, *J. Catal.*, **96**, 429 (1985)
[17] Y. Nitta, T. Utsumi, T. Imanaka and S. Teranishi, *Chem. Lett.*, 1399 (1984)
[18] Yu. I. Petrov and E.I. Klabunovskii, *Kinet. Catal.*, **8**, 814 (1967)
[19] E.I. Klabunovskii, A.A. Vedenyapin, E.S. Kapeiskaya, V.A. Pavlov and N.D. Zelvinskii, *Proc. 7th. Intern. Congr. Catal., Tokyo, 1980*, p. 390.
[20] T. Tanabe, O. Kazuo and Y. Izumi, *Bull Chem. Soc., Jpn.*, **46**, 514 (1973)
[21] J.A. van Dillen, J.W. Geus, L.A.M. Hermans and J. van der Meijden, *Proc. 6th. Intern. Congr. Catal., London, 1976*, p.677.
[22] S.D. Jackson, G.Kelly and G. Webb, unpublished results.
[23] W. Ramanowski, "Highly Dispersed Metals as Adsorbents" Wiley, New York, 1987 p. 171.
[24] M.A. Keane and G. Webb, *J. Mol. Catal.*, **73**, 91 (1992)
[25] G. Wittmann, G.B. Bartok, M. Bartok and G.V. Smith, *J. Mol. Catal.*, **60**, 1 (1990)

[26] I.M. Sutherland, A. Ibbotson, R.B. Moyes and P.B. Wells, *J. Catal.*, **125,** 77 (1990)

[27] T. Harada, K. Hanaki, Y. Izumi, J. Muaka, H. Ozaki and A. Tai, *Proc. 6th. Intern. Congr. Catal., London, 1976*, p. 1024.

[28] T. Harada, A. Tai, M. Yamamoto, H. Ozaki and Y. Izumi, *Proc. 7th. Intern. Congr. Catal., Tokyo, 1980*, p. 364

[29] T. Harada, M. Yamamoto, S. Onaka, H. Imaida, H. Ozaki, A. Tai and Y. Izumi, *Bull Chem. Soc. Jpn.*, **54,** 2823 (1981)

ENANTIO–DIFFERENTIATING HYDROGENATION OF SIMPLE ALKANONES WITH ASYMMETRICALLY MODIFIED HETEROGENEOUS CATALYST

Tsutomu Osawa [1], Akira Tai [2], Yoshimi Imachi [3], and Seiji Takasaki [3]

[1] Tottori University, Faculty of General Education
   Koyamacho–minami
   Tottori 680, Japan
[2] Himeji Institute of Technology, Faculty of Science
   Kanaji, Kamigori
   Hyogo 678–12, Japan
[3] Kawaken Fine Chemicals Co. Ltd.
   Imafuku, Kawagoe
   Saitama 350–11, Japan

ABSTRACT

The enantio–differentiating hydrogenation of 3–octanone was carried out using tartaric acid–NaBr–modified nickel catalyst and compared with that of 2–alkanones. For the hydrogenation of 3–octanone, fine nickel powder was a better source of the catalyst than Raney nickel. 3–Octanone was hydrogenated with an optical yield of 30% over tartaric acid–NaBr–fine nickel powder. As for the effect of hydrogenation temperature on optical yield, hydrogenation at a lower temperature resulted in a lower optical yield for the hydrogenation of 3–octanone, while the lower temperature resulted in a higher optical yield for the hydrogenation of 2–alkanones.

INTRODUCTION

The tartaric acid–NaBr–modified nickel catalyst (TA–NaBr–MNi) can be easily prepared by soaking a nickel catalyst in an aqueous solution of TA and NaBr. This simple catalyst gives around 80% optical yield in the hydrogenation of 2–alkanones by differentiating methyl and other alkyl groups attached to the carbonyl group of the substrate. For this promising catalyst, the hydrogenation of 3–alkanones to optically active 3–alkanols is the next challenging subject. To explore the differentiation ability of the catalyst between ethyl and pentyl groups, the hydrogenation of 3–octanone was carried out under various conditions and compared with that of the 2–alkanones.

Chiral Reactions in Heterogeneous Catalysis
Edited by G. Jannes and V. Dubois, Plenum Press, New York, 1995

EXPERIMENTAL

The GLC of the product was carried out using a Shimadzu GC–8A gas chromatograph. The optical rotations were measured using a JASCO DIP–370 polarimeter. The X–ray diffraction pattern of the catalyst was measured using a Rigaku RAD–B diffractometer. The mean nickel crystallite size was calculated from the half–width of the peak from the (111) plane of nickel metal.

Raney Nickel (RNi)

The Ni–Al alloy (Ni/Al=42/58, 1.9 g) was added to a 20% NaOH solution in small portions. The resulting suspension was held at 100 °C for 1 h. After the supernatant was removed by decantation, the catalyst was washed with twenty 25 ml portions of deionized water.

Fine Nickel Powder (FNiP)

Commercially available fine nickel powder (Vacuum Metallurgical Company Ltd., Chiba, Japan) with a mean particle diameter of 20 nm (1 g) was treated with a hydrogen stream for 0.5 h at the temperature stated in the text.

Modification

1) RNi was modified with a 100–ml solution containing 1 g of (R,R)–TA and 6 g of NaBr (pH of this solution had been adjusted to 3.2 with 1 mol/dm^3 NaOH) for 1 h at the temperature stated in the text, and 2) FNiP was modified with a 100–ml solution of (R,R)–TA (1 g) and NaBr (0.1 g) (pH 3.5) for 1 h at the temperature stated in the text.

Hydrogenation of Alkanones

TA–NaBr–MNi thus obtained was used for the hydrogenation of alkanones (4.1 g) in the mixture of pivalic acid (8.1 g) and tetrahydrofuran (THF) (10 ml) under an initial hydrogen pressure of 9 MPa. The hydrogenation temperature is stated in the text. When no further hydrogen consumption was observed, the catalyst was removed by decantation and THF was evaporated *in vacuo*. The hydrogenation product was dissolved in ether and washed with a saturated aqueous solution of K_2CO_3. The ether solution was then dried over anhydrous Na_2SO_4 and concentrated *in vacuo*. Simple distillation gave the product with a purity higher than 98% (GLC analyses: 5% Thermon 1000 on Chromosorb W at 75 °C (3–methyl–2–butanol) or at 70–260 °C (except 3–methyl–2–butanol)).

Determination of Optical Yield

The optical yield of the reaction was evaluated using the optical purity of the product determined by polarimetry. The specific optical rotations $[\alpha]_D^{20}$ of the optically pure enantiomers are: (S)–2–hexanol, $[\alpha]_D^{20} = +11.57°$ (neat) [1]; (S)–2–octanol, $[\alpha]_D^{20} = +9.76°$ (neat) [1]; (S)–2–decanol, $[\alpha]_D^{20} = +8.68°$ (neat) [1]; (S)–3–methyl–2–butanol, $[\alpha]_D^{20} = +4.85°$ (neat) [2]; and (S)–3–octanol, $[\alpha]_D^{20} = +8.22°$ (neat) [3].

RESULTS AND DISCUSSION

In the course of the studies of the enantio–differentiating hydrogenation of 2–alkanones, optical yield was found to be affected by both the catalyst preparation conditions and the hydrogenation conditions. These conditions were independently optimized to get high optical yield in the hydrogenation of 2–alkanones. The present study, where this catalyst is applied to the hydrogenation of 3–octanone, was effectively carried out using a similar approach.

Conditions of Catalyst Preparation

Tartaric acid–NaBr–modified RNi (TA–NaBr–MRNi) is an excellent catalyst for the enantio–differentiating hydrogenation of various 2–alkanones. An optical yield of about 80% was obtained with a series of straight chain 2–alkanones, and the highest optical yield of 85% was obtained during the hydrogenation of a branched 2–alkanone (3–methyl–2–butanone) at 60 °C (table 1). In spite of such performance with 2–alkanones, TA–NaBr–MRNi gave a low optical yield of around 10% for the hydrogenation of 3–octanone at 100 °C.

Table 1. Hydrogenation of 2–alkanones over (R,R)–TA–NaBr–MRNi [a].

No.	Substrate	Optical Yield / % (Configuration of the product)		
		120°C[b]	100°C[b]	60°C[b]
1	2-Hexanone	–	66(S)[d]	80(S)[d]
2	2-Octanone[c]	–	16(S)[e]	–
3	2-Octanone	50(S)[d]	66(S)[d]	80(S)[d]
4	2-Decanone	–	58(S)[d]	76(S)[d]
5	3-Methyl-2-butanone	–	63(S)[d]	85(S)

a) Modified with TA 1g and NaBr 6g at pH 3.2, 100°C b) Hydrogenation temperature

c) Hydrogenated over MRNi modified with TA 1 g and NaBr 4 g at pH 3.2, 0°C.

d) Reference (4) e) Reference (5)

For the hydrogenation of 3–octanone, TA–NaBr–MNi prepared from FNiP was much better than TA–NaBr–MRNi with respect to reproducibility and enantio–differentiating ability. FNiP was activated in a H_2 stream before modification and the temperature of this operation affected optical yield (figure 1). The treatment at higher temperature gave the higher optical yield and reached a maximal optical yield of about 30% at 320 °C. Figure 2 shows the effect of H_2–treatment time on optical yield at 200, 250, and 300 °C. Optical yield mainly depended on the temperature, and less on H_2–treatment time. Optical yield was almost constant from 30 min. to 60 min. of treatment. The catalyst was observed to be apparently aggregated after the H_2–treatment. In order to find the reason of increasing optical yield with higher H_2–treatment temperature, the mean nickel crystallite size was measured (table 2). The mean nickel crystallite size became larger with the elevation in H_2–treatment temperature and higher optical yield was obtained with such higher temperature treatment. Nitta *et al.* indicated that the catalyst with the larger crystallite size gave the higher optical yield in the enantio–differentiating hydrogenation of methyl acetoacetate [6]. They also predicted that the catalyst with a larger crystallite size had regularly–arranged nickel atoms on the catalyst surface providing sites for a strong and regular adsorption of the modifier, propitious to obtain high optical yield. When these facts were taken into account, our results indicated that a higher H_2–treatment temperature induced the larger FNiP crystallite size and it would provide the appropriate surface for the enantio–differentiating hydrogenation of 3–octanone.

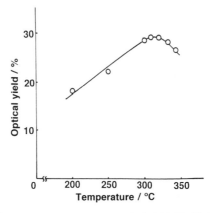

Figure 1. Effect of the H_2–treatment temperature on optical yield. Modifying temperature : 100 ℃, Hydrogenation temperature : 120 ℃.

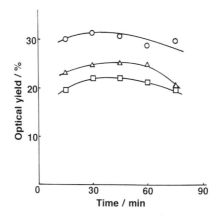

Figure 2. Effect of the H_2–treatment time on optical yield. Temperature of H_2–treatment : ○ = 300 ℃, △ = 250 ℃, □ = 200 ℃. Modifying temperature : 100 ℃, hydrogenation temperature : 120 ℃.

As for the modifying conditions, a higher modifying temperature is known to increase the optical yield of 2–alkanone hydrogenation over RNi [5], methyl acetoacetate over RNi [7] and reduced nickel catalyst [8]. Harada *et al.* revealed that the presence of aluminum or related metal compounds in a RNi catalyst was unfavorable for the effective enantio–differentiating catalyst and that high temperature modification removed the unfavorable ones from the catalyst surface [7]. Harada *et al.* also demonstrated that in the hydrogenation over a reduced nickel catalyst, high temperature modification was considered to remove the amorphous parts of the catalyst where racemic products were produced [8]. In the hydrogenation of 3–octanone over FNiP, the modifying temperature also affected the optical yield (table 3). The higher temperature modification improved the optical yield for the hydrogenation of 3–octanone over FNiP as well as that of 2–alkanones and methyl acetoacetate. Table 3 also shows the effect of modification temperature on the mean nickel crystallite size. Mean crystallite size slightly increased with the higher temperature modification, while the optical yield was significantly increased. These results indicate that the effect of high temperature modification is mainly attributed to the conditioning of the surface of the catalyst and not to the mean crystallite size of the catalyst. The conditioning of the FNiP catalyst surface was necessary to obtain high optical yield in spite of having no other metals such as aluminum which is unfavorable for the high enantio–differentiating ability of the catalyst. This conditioning process would remove amorphous parts from the surface and provide a smooth surface appropriate for the regularly–arranged adsorption of TA.

Table 2. Correlation of mean crystallite size with H_2–treatment temperature.

H_2-treatment temperature / °C	Optical yield / %[a]	Mean crystallite size / nm
---[b]	---	11.6
200	23	13.6
250	25	14.2
300	31	17.5

a) Obtained over TA-NaBr-MFNiP; modifying temperature : 100°C, hydrogenation temperature : 120°C, b) Without H_2 treatment and modification

Table 3. Effect of modifying temperature on mean crystallite size and optical yield.

H_2-treatment temperature / °C	Modifying temperature / °C	Crystallite size / nm	Optical yield / %[a]
300	100	17.5	31
300	0	16.7	12
200	100	13.0	23
200	0	12.6	10

a) Hydrogenation temperature : 120°C

Hydrogenation Conditions

For the hydrogenation of 3–octanone as well as 2–alkanones, the addition of pivalic acid to the reaction system was indispensable to obtain a high optical yield. However, the effect of hydrogenation temperature on optical yield was different for both reactions (tables 1 and 4). Hydrogenation at a lower temperature resulted in a lower optical yield for the hydrogenation of 3–octanone, while the lower temperature resulted in a higher optical yield for the hydrogenation of 2–alkanones.

The present study revealed that 3–octanone was hydrogenated with an optical yield of 30% over TA–NaBr–MFNiP. The enantio-differentiating hydrogenation of 3–octanone over homogeneous and heterogeneous catalyst has never been studied before. So this study shows TA–NaBr–MNi is a promising catalyst for the enantio–differentiating hydrogenation of 3–alkanones. Ethyl and pentyl groups (3–octanone) were much more difficult to be differentiated from each other as compared to methyl and other alkyl groups (2–alkanones) over TA–NaBr–MNi. A high degree of catalyst enantio–differentiating ability was required to distinguish between ethyl and pentyl groups, because the difference in the size of alkyl groups in 3–octanone was much smaller than that in 2–alkanones. We have proposed an enantio–differentiating model for the hydrogenation of 2–alkanones (figure 3) [9].

Table 4. Hydrogenation of 3–octanone over (R,R)–TA–NaBr–MNi.

No.	Catalyst	Optical Yield / % (Configuration of the product)		
		120°C[a)]	100°C[a)]	60°C[a)]
1	RNi[b)]	9(S)	10(S)	3(R)
2	FNiP[c)]	13(S)	11(S)	3(S)
3	FNiP[d)]	25(S)	26(S)	15(S)

a) Hydrogenation temperature b) Modified with TA 1g and NaBr 1g at pH 3.2 100°C
c) H$_2$ treatment at 200°C, modified with TA 1 g and NaBr 50 mg at pH 4.0, 0°C.
d) H$_2$ treatment at 200°C, modified with TA 1 g and NaBr 0.1 g at pH 3.5, 100°C.

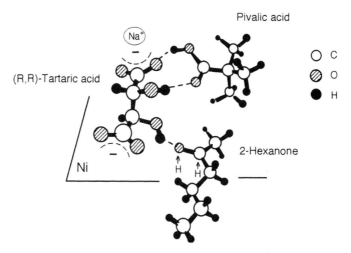

Figure 3. Mode of enantio–differentiation in the hydrogenation of 2–alkanones.

Differentiation between methyl and other alkyl groups was carried out along the following steps : 1) TA on the catalyst surface forms an associative complex with pivalic acid added to the reaction system, 2) the substrate is fixed to the catalyst surface through TA, 3) the TA–pivalic acid complex recognizes the structure of the substrate through interaction between the alkyl group of pivalic acid and that of 2–alkanones, and 4) hydrogen was added to the substrate from the catalyst surface. According to this model, carboxylic acid added to the reaction system was the key factor for creating the enantio–differentiating ability of this catalyst in the hydrogenation of 2–alkanones. Therefore, more detailed comparative studies of the enantio–differentiating hydrogenation of 2– and 3–alkanones including the structure of the carboxylic acid added to the reaction system would lead to the development of a highly efficient heterogeneous catalyst for the enantio–differentiating hydrogenation of 3–alkanones and also the elucidation of the necessary factors to differentiate between various alkyl groups.

ACKNOWLEDGEMENTS

The authors thank Professor Tadao Harada of Ryukoku University for many helpful discussions and suggestions during the course of this work.

REFERENCES

[1] R. H. Richard and J. Kenyon, *J. Chem. Soc.*, **99**, 45 (1911).
[2] R. H. Richard and J. Kenyon, *J. Chem. Soc.*, **101**, 620 (1912).
[3] R. H. Richard and J. Kenyon, *J. Chem. Soc.*, **103**, 1923, (1913).
[4] T. Osawa, T. Harada, and A. Tai, *J. Catal.*, **121**, 7 (1990).
[5] T. Osawa and T. Harada, *Bull. Chem. Soc. Jpn.*, **57**, 1518 (1984).
[6] Y. Nitta, F. Sekine, T. Imanaka, and S. Teranishi, *Bull. Chem. Soc. Jpn.*, **54**, 980 (1981).
[7] T. Harada, M. Yamamoto, S. Onaka, M. Imaida, H. Ozaki, A. Tai, and Y. Izumi, *Bull. Chem. Soc. Jpn.*, **54**, 2323 (1981).
[8] T. Harada, Y. Imachi, A. Tai, and Y. Izumi, Proceedings on Metal–support and Metal–additive Effects in Catalysis, Lyon, p.377, 1982.
[9] T. Osawa, T. Harada, and A. Tai, Abstracts of the 7th International Symposium on Relations Between Homogeneous and Heterogeneous Catalysis, Tokyo, p.230, 1992.

ENANTIO–DIFFERENTIATING HYDROGENATION OF 2–BUTANONE : DISTINCTION BETWEEN CH3 AND C2H5 WITH A MODIFIED NICKEL CATALYST

Tadao Harada[1] and Tsutomu Osawa[2]

[1] Ryukoku University, Faculty of Science and Technology, Seta, Otsu 520–21, Japan
[2] Tottori University, Faculty of General Education, Koyamacho–minami, Tottori 680, Japan

ABSTRACT

The enantio–differentiating hydrogenation of 2–butanone was carried out using Raney nickel modified with tartaric acid and sodium bromide. 2–Butanol of 72% optical purity was obtained by optimizing the reaction variables and modification variables. The degree of the intrinsic enantio–differentiating ability of the adsorbed tartaric acid for 2–butanone was supposed to be similar to the abilities of tartaric acid for higher 2–alkanones.

INTRODUCTION

The tartaric acid–sodium bromide–modified Raney nickel catalyst (TA–NaBr–MRNi) is an excellent heterogeneous catalyst for the enantio–differentiating hydrogenation of various ketones [1]. TA–NaBr–MRNi is the only effective catalyst for the enantiodifferentiating hydrogenation of 2–alkanones. When the (R,R)–isomer of tartaric acid (TA) was used as a modifier of the Raney nickel catalyst (RNi), (S)–2–alkanols were preferentially obtained. An optical yield (OY) of 70–80% was attained from the hydrogenation of straight chain 2–alkanones longer than 2–pentanone or α–branched 2–alkanones with TA–NaBr–MRNi [1,2]. However, 2–butanone, which requires the catalyst to distinguish between CH3 and C2H5 groups, was hydrogenated at a rather low OY (58–63%). This fact shows the difficulty of a one–carbon distinction by this catalyst. In this report, we wish to present the results of our effort for improving the OY during the hydrogenation of 2–butanone over TA–NaBr–MRNi.

EXPERIMENTAL

Typical experimental conditions are as follows [1,3] :

Raney Nickel

The Ni–Al alloy (Ni/Al=42/58, the amount stated in the text) was added to an aqueous solution of NaOH (the weight of NaOH : 3 times that of the alloy) in small portions. The resulting suspension was held at 100 °C for 1 h. After the supernatant was removed by decantation, the catalyst was washed 15–20 times with deionized water to remove most of the NaOH.

Modification of Raney Nickel

The RNi thus obtained was modified with an aqueous solution of TA and NaBr. The amounts of TA, NaBr, and water in the solution are stated in the text. The pH value of this solution was adjusted to 3.2 with an aqueous solution of NaOH. The modification of RNi was carried out at 100 °C for 1h. After removal of the solution by decantation, the resulting catalyst was successively washed with water, methanol, and tetrahydrofuran (THF).

Hydrogenation of 2–butanone

The TA–NaBr–MRNi thus obtained was employed for the hydrogenation of 2–butanone. The reaction was carried out in a mixture of pivalic acid and THF under an initial hydrogenation pressure of 9 MPa until no further consumption of hydrogen was observed. The amounts of 2–butanone, pivalic acid, and THF are stated in the text.

Determination of OY

After removal of the catalyst by decantation and removal of THF by evaporation, the hydrogenation product was dissolved in ether and washed with a saturated aqueous solution of K_2CO_3. The ether solution was then dried over anhydrous Na_2SO_4 and concentrated in vacuo. 2–Butanol (chemical purity, 95–98%), contaminated with the solvents, was obtained after a simple distillation. Some parts of the distilled 2–butanol were further purified by preparative GLC (20% PEG 20M on Chromosorb W at 100 °C) for the determination of OY. The OY value was determined based on the optical purity of the product. The optical purity was estimated by polarimetry.

RESULTS

The reported OY value for the hydrogenation of 2–butanone [1,2] was attained under the optimal conditions for the hydrogenation of 2–octanone (NaBr/TA ratio in the modi-fying solution : 8.8 (mol/mol), pivalic acid/2–octanone ratio in the reaction mixture : 1.7 (mol/mol), reaction temperature : 60 °C). The improvement in OY for the hydrogenation of 2–butanone was attempted by revising the reaction and modification variables.

Reaction Variables

The hydrogenation of 2–butanone was carried out under various operating conditions. The catalyst was prepared under the most efficient conditions providing for the hydro–genation of 2–octanone. The number of reaction variables affecting the OY are considerable. In the course of this study, we investigated the relation between the OY and reaction variables, such as reaction solvent and carboxylic acid in the reaction mixture, reaction temperature, and the amount of TA–NaBr–MRNi. We could not succeed in finding better reaction conditions than the optimal conditions for 2–octanone. The OY value decreased or did not change by changing the variables. Some results are shown in tables 1 and 2, and figure 1. Table 1 shows the relation between the OY and the reaction temperature. The optimal reaction temperature was 50–60 °C. Table 2 shows the effect of added carboxylic acid on the OY. As in the hydrogenation of 2–octanone, pivalic acid was the best carboxylic acid additive. Figure 1 shows the relation between the amount of pivalic acid in the reaction mixture and the resulting OY. The addition of more than one mol–equivalent of pivalic acid to 2–butanone was indispensable for attaining a high OY. The optimal reaction conditions for the hydrogenation of 10.0 g (140 mmol) of 2–butanone were found to be as follows. TA–NaBr–MRNi : prepared from 3–20 g of the RNi alloy (Ni/Al=42/58). Added carboxylic acid : pivalic acid (>14.2 g (140 mmol)). Reaction solvent : THF (20 ml). Reaction temperature : 50–60 °C.

Table 1. Relation between optical yield and reaction temperature[a]

Reaction Temperature (°C)	Optical Yield (%)
70	54
60	60
55	59
50	60
40	56

(a) MRNi : the RNi prepared from the RNi alloy (3.8 g) was modified with TA (2.0 g, 13 mmol) and NaBr (12 g, 120 mmol) at pH 3.2, 100 °C. Reaction mixture : 2–butanone (10 g, 140 mmol) and pivalic acid (18 g, 180 mmol).

Modification Variables

The RNi was modified under various conditions, and the resulting TA–NaBr–MRNi was used under the best reaction conditions providing for the hydrogenation of 2–butanone. Table 3 shows the relation between the OY and the NaBr/TA ratio (mol/mol) in the modifying solution. When the ratio was 22, the OY reached its maximal value. Figure 2 shows the relation between the OY and the pH value of the modifying solution. The modification at pH 3.2–3.3 was the best for the TA–NaBr–MRNi preparation as in the case of the catalyst for 2–octanone hydrogenation. The tartaric acid in the modifying solution functions not only as a chiral adsorbate on the catalyst surface, but also as a reagent for reducing the aluminum–contaminated surface area, where the production of racemic 2–butanol can be assumed (non–enantio–differentiating area) [3,4]. Thus, the amount of the modifying solution was supposed to affect the OY. Figure 3 shows the relation between the amount of modifying solution and the resulting OY. The addition of more than 300 ml of modifying solution to the RNi prepared from 10.0 g of the RNi alloy was indispensable for attaining a high OY.

Table 2. Effect of added carboxylic acid on optical yield [a]

Carboxylic acid	Ratio to 2–butanone (mol/mol)	Optical yield (%)
none	0.00	9
Acetic acid	0.48	48
Propionic acid	0.65	52
n–Butyric acid	0.65	52
iso–Butyric acid	0.16	42
	0.33	52
	0.65	56
	0.98	52
	1.47	16
iso–Valeric acid	0.65	50
Pivalic acid	0.21	48
	1.70	60

(a) MRNi : the RNi prepared from the RNi alloy (3.8 g) was modified with TA (2.0 g, 13 mmol) and NaBr (12 g, 120 mmol) at pH 3.2, 100 °C. Reaction mixture : 2–butanone (10 g, 140 mmol) and the stated amount of carboxylic acid. Reaction temperature : 60 °C.

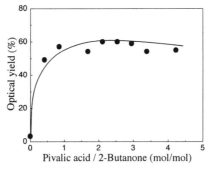

Figure 1. Relation between optical yield and pivalic acid/2–butanone ratio.
MRNi : the RNi prepared from the RNi alloy (20 g) was modified with TA (10 g, 67 mmol) and NaBr (60 g, 590 mmol) at pH 3.2, 100 °C. Reaction mixture : 2–butanone (10 g, 140 mmol), pivalic acid (0–60 g, 0–590 mmol), and THF (20 ml). Reaction temperature : 50 °C.

The hydrogenation product was esterified with 3,5–dinitrobenzoyl chloride. The crude ester became optically pure after five recrystallizations from methanol. The specific optical rotation ($[\alpha]_D^{20}$) of the (S)–ester was +30.7 (c 5, CHCl$_3$).

Table 3. Relation between optical yield and NaBr/TA ratio in modifying solution [a]

NaBr/TA (mol/mol)	Optical Yield (%)
7.4	59
8.8	60
14.7	69
22.1	72
29.4	66
36.8	54

(a) MRNi : the RNi prepared from the RNi alloy (10 g) was modified with an aqueous
solution containing TA (4.0 g, 27 mmol) and NaBr at pH 3.2, 100 °C. Reaction
mixture : 2–butanone (5 g, 70 mmol), pivalic acid (18 g, 180 mmol), and THF (20 ml).
Reaction temperature : 50 °C.

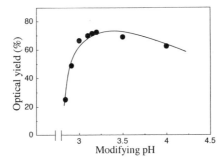

Figure 2. Relation between optical yield and modifying pH.
MRNi : the RNi prepared from the RNi alloy (10 g) was modified with TA (4.0 g, 27 mmol) and NaBr (60 g,
590 mmol) at 100 °C. Reaction mixture : 2–butanone (5.0 g, 70 mmol), pivalic acid (18 g, 180 mmol), and
THF (20 ml). Reaction temperature : 50 °C.

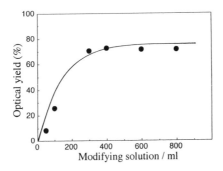

Figure 3. Relation between optical yield and amount of modifying solution.
MRNi : the RNi prepared from the RNi alloy (10 g) was modified with an aqueous solution of TA and NaBr at
pH 3.2, 100 °C. 100 ml of modifying solution contained 1.0 g (6.7 mmol) of TA and 15 g (150 mmol) of NaBr.
Reaction mixture : 2–butanone (5.0 g, 70 mmol), pivalic acid (18 g, 180 mmol) and THF (20 ml). Reaction
temperature : 50 °C.

DISCUSSION

The roles of the adsorbed TA and NaBr have been interpreted as follows [4]. There exists an enantio–differentiating area (e.d. area) and a non–enantio–differentiating area (non–e.d. area) on the RNi. By modification with TA, the e.d. area adsorbs TA and yields optically active products with the aid of TA. The non–e.d. area does not adsorb TA and yields only racemic products. When an aqueous solution of TA and NaBr is used as the modifying solution, the TA and the NaBr are adsorbed on the e.d. area and the non–e.d. area, respectively. The adsorbed NaBr reduces the hydrogenation activity of the non–e.d. area to improve the OY. The OY value (72%) attained in this study suggests that the intrinsic enantio–differentiating ability of the adsorbed TA for 2–butanone is similar to the abilities of TA for the other 2–alkanones.

It has been believed that efficient enantio–differentiation is possible only when more than two strong interactions (hydrogen bondings, dipole–dipole interactions, etc.) are formed between the substrate and enantio–differentiating reagent or catalyst. Alkanones can form only a strong interaction with enantio–differentiating reagent or catalyst through its carbonyl group. Moreover, the enantio–differentiating hydrogenation of the simplest prochiral alkanone, 2–butanone, requires a one–carbon distinction by the reagent or catalyst. Thus, the study of the enantio–differentiating hydrogenation of 2–butanone is a challenging project in the field of studying the enantio–differentiating reactions. The OY value (72%) attained in this study is outstanding when the fact is taken into account that a low ee(48%) was reported for the hydrogenation of 2–butanone with alcohol dehydrogenase from *Thermoanaerobium brockii* [6]. An investigation is now in progress for development of a more efficient catalyst for the enantio–differentiating hydrogenation of 2–butanone.

ACKNOWLEDGEMENTS

The authors thank Professors Yoshiharu Izumi and Akira Tai for their helpful suggestions.

REFERENCES

[1] T. Osawa, T. Harada, and A. Tai, *J. Catal.*, **121**, 7 (1990).
[2] T. Osawa, *Chem. Lett.*, 1609 (1986).
[3] T. Harada, M. Yamamoto, S. Onaka, M. Imaida, H. Ozaki, A. Tai, and Y. Izumi, *Bull. Chem. Soc. Jpn.*, **54**, 2323 (1981).
[4] T. Harada, A. Tai, M. Yamamoto, H. Ozaki, and Y. Izumi, *Proc. 7th Int. Congr. Catal. Tokyo,* 364 (1980).
[5] T. Osawa and T. Harada, *Bull. Chem. Soc. Jpn.*, **57**, 1518 (1984).
[6] E. Keinan, E.K. Hafeli, K.K. Steh, and R. Lamed, *J. Am. Chem. Soc.,* **108**, 162 (1986).

CHIRAL HYDROGENATION OF ESTRONE AND ESTRONE-3-METHYL ETHER

Gy. Göndös[1], Gy. Wittmann[1], M. Bartók[1] and J.C. Orr[2]

[1] József Attila University
Department of Organic Chemistry
Dóm tér 8
H–6720 Szeged, Hungary

[2] Memorial University of Newfoundland
Faculty of Medicine
St. John's, NF, A1B 3V6, Canada

INTRODUCTION

We have previously studied the use of enantiomeric chiral reducing agents in solution to control the direction of the hydrogenation of the keto group of estrone and estrone–3–methyl ether [1]. The complex catalysts chiral hydrosilane–rhodium–(+)– or (–)–DIOP or (+)– or (–)–BINAP allowed different stereoselectivities in the formation of the 17α– and 17 β–alcohols.

The hydrogenation of steroid ketones under heterogeneous catalytic conditions over Raney nickel (RNi) catalyst has been described [2], but chiral RNi catalysts have not been explored for the asymmetric hydrogenation of steroid ketones.

The present communication reports results on the asymmetric reduction on estrone–3–methyl ether with (2R,3R)–(+)– or (2S,3S)–(–)–tartaric acid–modified RNi (TA–RNi) as catalyst, as compared with the same reaction under homogeneous catalytic conditions.

The influence of the two chiral tartaric acid enantiomers on the keto group reduction was studied as a function of pH and of the hydrogen pressure.

EXPERIMENTAL

1. Homogeneous hydrogenation : In one analytical procedure, 10 μmol estrone or estrone–3–methyl ether in benzene or xylene, under argon, in the presence of 20 μmol hydrosilane, was allowed to react with (+)– or (–)–DIOP or (+)– or (–)–BINAP–Rh/S/Cl

Chiral Reactions in Heterogeneous Catalysis
Edited by G. Jannes and V. Dubois, Plenum Press, New York, 1995

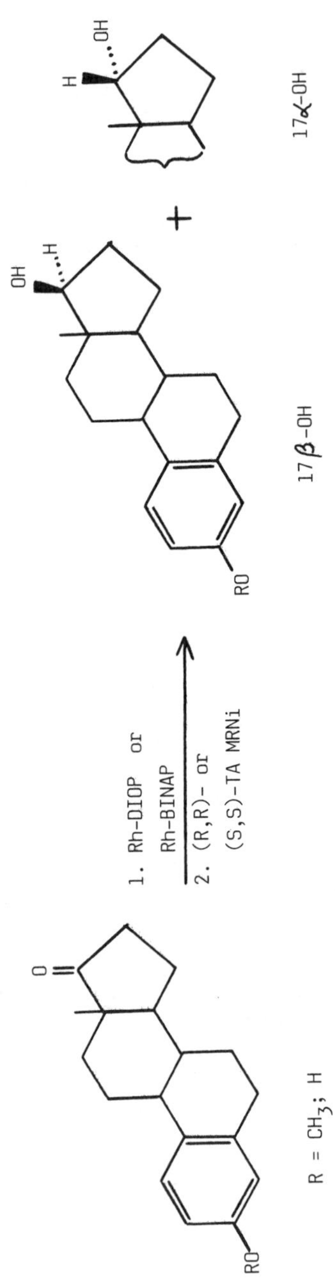

Scheme 1.

(0.0305 or 0.061 μmol, S = solvent) prepared in situ from [(cyclooocta–1,5–diene)RhCl]$_2$. The reaction was carried out for various times and at various temperatures. The steroid silyl ethers obtained were hydrolysed with a 0.1% methanolic solution of p–toluenesulphonic acid to give the corresponding free alcohols. The substrate and the products were separated by means of HPLC on a Partisil PXS 10/25 column (Whatman) [3]. UV measurements at an appropriate wavelength, with a simultaneous refractive index procedure, were used for detection.

2. Heterogeneous hydrogenation : RNi was prepared from Ni/Al alloy (Fluka : Ni/Al 1:1) as described earlier [4]. The hydrogenation of ethyl acetoacetate was used to characterize the prepared catalyst.

In a hydrogenation vessel, the substrate (0.879 mmol) in 15 ml abs. THF was hydrogenated over the given molar ratio (2.0489–3.4822) of catalyst under 1 or 80 atm. initial hydrogen pressure at 40 °C for 8 hours.

The aluminium and nickel contents of the filtered catalyst were determined by the titrimetric method previously described [5]. The hydrogenated products were analysed by HPLC as described above.

RESULTS AND DISCUSSION

Under homogeneous catalytic conditions, with the complex catalyst (+)–Rh–DIOP, the 17ß–hydroxy isomer was observed in large excess (17ß:17α = 83:17) whereas with (–)–Rh–DIOP there was a mild preponderance of the 17α–hydroxy isomer (17ß:17α = 46:54). Estrone–3–methyl ether with Rh–(+)–DIOP in benzene at 22 °C gave essentially the same ratio (82:18). Substitution of Rh–(–)–DIOP for the (+) enantiomer gave a ratio of 60:40 for estrone–3–methyl ether.

Table 1. Hydrosilylation[a] of estrone and estrone–3–methyl ether by chiral Rh–DIOP or Rh–BINAP catalysts.

Substrate	Catalyst	Conversion(%)	Configuration		Reference
			17ß-OH(%)	17α-OH(%)	
Estrone[b]	(+)-DIOP-Rh	79	83	17	(1) a,b
	(-)-DIOP-Rh	89	46	54	
Estrone-3-methyl ether[b]	(+)-DIOP-Rh	38	82	18	(1) a,b
	(-)-DIOP-Rh	34	60	40	
Estrone[c]	(+)-BINAP-Rh	44	77	23	
	(-)-BINAP-Rh	84	83	17	
Estrone-3-methyl ether[c]	(+)-BINAP-Rh	96	34	66	
	(-)-BINAP-Rh	98	93	7	

[a] Hydrosylane was: diphenyl dihydrosylane

[b] Solvent was: benzene

[c] Solvent was: xylene

Table 2. Hydrogenation of estrone-3-methyl ether over modified Raney-Ni Catalysts

Catalyst	Modifying pH	H$_2$ pressure (atm)	Substrate/cat mmol/g	Conversion (%)	Configuration 17β-OH(%)	17α-OH(%)
(2S,3S)-(−)-TARNi[a]	7.0	1	3.48	48	67	33
(2R,3R)-(+)-TARNi	7.0	1	1.48	66	71	29
(2S,3S)-(−)-TARNi	3.5	1	2.05	26	60	40
(2R,3R)-(+)-TARNi	3,5	1	2.53	16	70	30
(2S,3S)-(−)-TARNi	10.0	1	3.48	38	73	27
(2R,3R)-(+)-TARNi	10.0	1	2.76	22	78	22
(2R,3R)-(+)-TARNi	3.5	80[b]	3.09	47	83	17
RNi	7.0	1	2.45	97	73	27

[a] TARNi: Raney nickel modified with chiral tartaric acid.

[b] In the high pressure hydrogenation, the reaction mixture was transferred to a stainless steel autoclave and shaken continuously for 8 hours at 40 °C at an initial hydrogen pressure of 80 atm.

Reduction of estrone with diphenyldihydrosilane and Rh–(+)–BINAP in xylene at 22 °C gave a 17ß– : 17α–alcohol ratio of 77:23. However, estrone–3–methyl ether in xylene at 22 °C gave predominantly the 17α–alcohol (66%). This is much better than for the methods previously available. Reduction of estrone and estrone–3–methyl ether with (–)– BINAP–Rh or diphenyldihydrosilane gave a 17ß:17α–alcohol ratio of 83:17 or 93:7 (table 1).

At pH 7, RNi catalysts with (2S,3S)–(–)–TA (17ß:17α = 67:33) and with (2R,3R)–(+)– TA (17ß:17α = 71:29) gave results that were little different from those for the unmodified catalyst (17ß:17α = 73:27). (table 2). The highest 17α:17ß alcohol ratio was obtained at pH 3.5, with (2S,3S)–(–)–TA–modified RNi catalyst, 40% 17α–alcohol being formed. The opposite chirality of the modifying reagent gave much less 17α–alcohol (30%). At pH 10, the 17α:17ß isomer ratios were lower than those obtained at pH 3.5 or pH 7. At all studied pH values, SS–(–)–TA–RNi gave a higher 17α:17ß ratio than did RR–(+)–TA–RNi. An increase in pressure, studied only with RR–(+)–TA–RNi at pH 3.5, gave the lowest 17α:17ß ratio yet found (83:17).

CONCLUSION

The complex catalysts Rh–DIOP and Rh–BINAP with opposite chiralities induce opposite asymmetries in the hydrosilylation of the prochiral molecule of estrone and estrone–3–methyl ether in the presence of diphenylhydrosilane. The highest 17α–hydroxy isomer yield was attained under conditions of homogeneous hydrogenation of estrone and estrone–3–methyl ether. Rh–(–)–DIOP and Rh–(+)–BINAP are useful mild reagents for the reduction of a 17–ketone to the relatively inaccessible 17α–alcohol.

The results of hydrogenation on a modified RNi catalyst revealed that the chirality of the modifier, the pH and the hydrogenation pressure are important factors governing the ratio of 17α– to 17ß–estradiol–3–methyl ether formed.

ACKNOWLEDGEMENT

We acknowledge support for this research provided by Hungarian National Science Foundation through grant OTKA 1885/1991.

REFERENCES

[1] a) Gy. Göndös and J.C. Orr, J.C.S. Chem. Commun., 1238 (1982).
     b) Gy. Göndös, L. Gera, M. Bartók and J. C. Orr, J. Organomet. Chem., 373, 365 (1989).
[2] a) J. Fried and J.A. Edwards (Eds), Organic Reactions in Steroid Chemistry, vol. I, Van Nostrand
        Reinhold, New York, p. 135 and pp 61–110, (1972).
     b) M. Bartók and K. Felföldi, Stereochemistry of Heterogeneous Metal Catalysis; John Wiley, Chichester,
        pp 360–371 (1985).
[3] Gy. Göndös and J. C. Orr in H. Kalász and L. S. Ettre (Eds),. Chromatography, the state of the art,
        Budapest, Akadémiai Kiadó, pp 113–117 (1985).
[4] a) T. Harada, Bull. Chem. Soc. Jpn., 53, 1019 (1980).
     b) A. Hoek, H.M. Woerde, W. M. H. Sachtler, in New Horizons in Catalysis, T. Seiyama and K. Tanabe,
        (Eds) Elsevier, Kodasha, Amsterdam, p. 376 (1981).
     c) Y. Izumi, Adv. Catal., 32, 215 (1984)
[5] M. Bartók, Gy. Wittmann, Gy. Göndös and G. V. Smith, J. Org. Chem., 52, 1139 (1987).

ENANTIOSELECTIVE HYDROGENATION OF α–KETOESTERS: A MOLECULAR VIEW ON THE ENANTIO–DIFFERENTIATION

A. Baiker[1], T. Mallat[1], B. Minder[1], O. Schwalm[2], K. E. Simons[1] and J. Weber[2]

[1]Department of Chemical Engineering and Industrial Chemistry
Swiss Federal Institute of Technology, ETH Zentrum
CH – 8092 Zürich, Switzerland, fax: (41–1) 632 11 63.
[2]Department of Physical Chemistry
University of Geneva
30 quai Ernest–Ansermet
CH – 1211 Geneva 4, Switzerland

ABSTRACT

Three mechanistic models, which have been proposed for the interpretation of enantioselective hydrogenation of α–ketoesters on platinum metal catalysts modified by cinchona alkaloids, are critically reviewed and compared. The characteristics and the implicit contradictions of the models are discussed in the light of relevant experimental observations.

INTRODUCTION

Chiral heterogeneous catalysis is a subject of growing interest in both the academic and industrial community. This interest arises from increasing requirements for compounds to be sold as their pure enantiomer, as opposed to racemic mixtures. Quite often different enantiomers have opposing properties, with only one enantiomer being active for the desired application, the other enantiomer being at best ballast and in some instances has undesired side–effects.

The number of known enantioselective systems is growing slowly, however, according to recent reviews [1–4] there are only two which have been extensively studied. These being the asymmetric hydrogenation of β–ketoesters over Ni catalysts modified by tartaric acid and α–ketoester hydrogenation over Pt catalysts modified by cinchona alkaloids. The latter reaction was first reported by Orito *et al.* [5–8] and has been studied further by several groups [9–12]. Most investigations have been carried out using the liquid phase hydrogenation of methyl– or ethyl pyruvate. The characteristic features of the reaction are depicted in scheme 1. Frequently used solvents are ethanol, toluene and acetic acid.

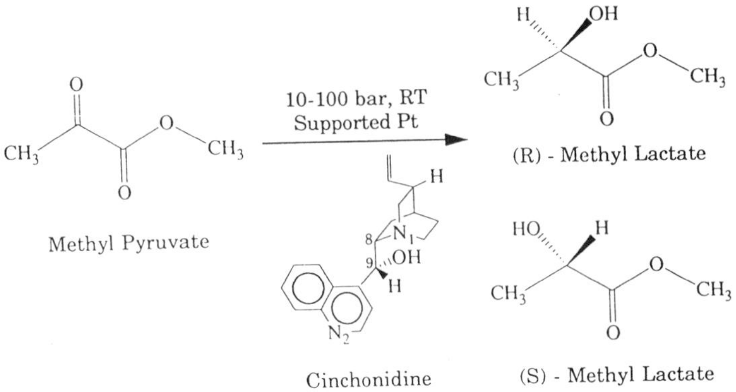

Scheme 1.

Any proposed mechanism for α–ketoester hydrogenation over cinchona–modified Pt metals must be able to explain several features of the reaction which have been found in the past years. The most important aspects are:

(i) the high degree of enantio–differentiation;

(ii) the five to one hundred–fold increase in reaction rate over that observed for the racemic reaction in the absence of modifier [9];

(iii) the loss of both enantioselectivity and activity after alkylation of the nitrogen of the quinuclidine ring (N_1), while the role of the quinoline ring and the interaction of pyruvate with the oxygen in the C_9–OH group of cinchonidine seems to be of secondary importance [13];

(iv) a change in the chirality of the stereogenic centres at the C_8 and C_9 positions of the alkaloid results in product of opposite enantiomeric excess [7,14];

(v) the requirement of large Pt particles of 3 nm diameter or more to achieve high enantiomeric excess [10,15] and

(vi) the influence of solvent in the reaction [5,6,16,17].

The first attempts to rationalise the reaction mechanism were made based on measurements of reaction kinetics. Kinetic models for quantifying the marked increase in the reaction rate occurring upon modification have been proposed by Wehrli [18], and Garland and Blaser [19]. These models, which are essentially similar, are based on a two–cycle mechanism as predicted for a ligand accelerated reaction where a slow unselective (unmodified catalyst) and a fast enantioselective reaction cycle are assumed to be in equilibrium.

A molecular view of the enantio–differentiation of the system was first reported by Wells and co–workers [11] in 1990. The authors suggested that the enantio–differentiation originates from a non close–packed array of cinchonidine molecules pre–adsorbed by their quinoline rings onto a Pt (100) surface in such a manner that shaped ensembles of Pt atoms were left exposed, onto which a methyl pyruvate molecule could conveniently only adsorb in a manner to give R–(+)–methyl lactate on hydrogenation. Adsorption of the pyruvate in a manner to give S–(–)–product was not possible. The argument for this was proposed on the basis that planar adsorption of the pyruvate onto the catalyst with the carbonyl groups orientated *top–left, bottom–right* relative to the central C–C bond would produce R–(+)–

methyl lactate. Alternatively, adsorption of the molecule with the carbonyl groups *bottom–left, top right* relative to the central C–C bond resulted in S–(–)–methyl lactate on hydrogenation. Although this argument accounted for enantio–differentiation, the carbonyl group to be hydrogenated was far from N_1, the importance of which was discovered soon after.

In the light of this finding the "template model" was further developed [3] and the reacting molecule positioned such that the carbonyl group was in the proximity of N_1. Two active sites were proposed and enantioselectivity was again explained in terms of steric hindrance: the pyruvate molecule could easily adsorb in a planar fashion to produce R–(+)–product on hydrogenation (as depicted in figure 1) but adsorption of the alternative enantioface of the pyruvate was hindered by a second alkaloid molecule, arbitrarily placed adjacent to the first. Furthermore, the half–hydrogenated state of the product was stabilised by a hydrogen bond to the alicyclic nitrogen (N_1) [20]. The second type of active site for enantioselective hydrogenation occurred on the edge of the Pt crystallite, where adsorption of a pyruvate molecule adjacent to the alkaloid was only possible in a manner which produced R–(+)–lactate on hydrogenation.

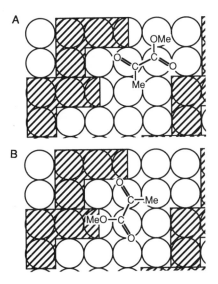

Figure 1. "Template model" : cinchonidine molecules adsorbed on Pt (100) forming a site in which methyl pyruvate can adsorb to produce R–(+)–lactate (A), but cannot adsorb to produce S–(–)–lactate (B) due to steric hindrance.

Another mechanism was proposed recently by Augustine *et al.* [12] to account for a phenomenon they observed whereby S–(–)–lactate formation could be achieved over Pt/alumina catalysts with low concentrations of cinchonidine. It was proposed that two different modes of adsorption of cinchonidine, at two different sites on a Pt surface could occur. When cinchonidine adsorbed flat through its quinoline ring system, at sites adjacent to corner atoms of a Pt crystallite (figure 2, A and B), a 1:1 complex with the pyruvate could be formed. This adduct, stabilised by a hydrogen bond between the C_9–OH and ester oxygen was orientated such that hydrogen was available only from the Pt corner atom located between the two moieties. No acceleration in reaction rate would result from this geometry which produced only S–(–)–lactate product.

For production of R–(+)–lactate, cinchonidine adsorption is considered to occur through N_2 with the quinoline ring perpendicular to the metal surface and adjacent to a Pt adatom (figure 2, C and D). A 1:1 adduct is again formed between the pyruvate and cinchonidine, but this time comprising a six membered ring formed by nucleophilic attraction between the quinuclidine nitrogen N_1 and the keto carbon, and the oxygen at C_9–OH and the ester oxygen. The formation of this species also accounted for the observed rate enhancement. The pyruvate molecule co–adsorbs with hydrogen onto the Pt adatom. As only one enantioface of the pyruvate molecule has access to the hydrogen from the adatom, R–(+) product results.

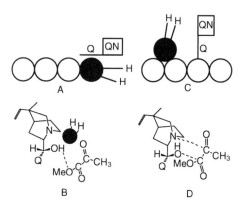

Figure 2. Augustine's model : two modes of adsorption of cinchonidine represented in figures A and C, and the structure of the corresponding transition complexes shown in figures B and D (Q : quinoline moiety, QN : quinuclidine moiety).

THEORETICAL CALCULATIONS

We have reported very recently [21] a theoretical study aimed at rationalising the interaction between the chiral modifier and the pyruvate ester. Quantum chemistry techniques at both semiempirical and *ab initio* levels and molecular mechanics have been used for this purpose. In order to involve indirectly the role of protic solvents, we investigated the possible interaction of pyruvate with both NH_3 and NH_4^+ species, which were chosen as simplified models of protonated and unprotonated cinchonidine [22].

The zone of highest reactivity for nucleophilic attack was found to be located between the ester– and keto–carbonyl groups of pyruvate, whereas the most favourable sites for electrophilic attack lay approximately in the regions of the lone pairs of the oxygen atoms. The pyruvate – NH_4^+ system was found to be much more stable (by 25 kcal mol^{-1}) due to favourable electrostatic interaction. This is an indication that in cases where cinchonidine can be protonated, this species will interact with the pyruvate. The other two possible binding sites of cinchonidine for electrophilic attack were not involved in the calculations, as:

(i) the basicity of N_2 nitrogen in the quinoline ring is lower by orders of magnitude than that of the tertiary nitrogen N_1 in the quinuclidine part;

(ii) changes at N_1 (alkylation) were found to be detrimental to the enantio–differentiation, while substitution of the OH group by OCH_3 or H had moderate influence [13].

Molecular mechanics using the Amber force field was used to optimize the possible conformations of the pyruvate – protonated cinchonidine system. The complex shown in figure 3A, which is proposed to be the precursor to R–(+)–methyl lactate, can be adsorbed in a planar π–bonding mode on the (flat) Pt surface via the aromatic quinoline ring, without hindering the interaction of the carbonyl groups of methyl pyruvate with the Pt surface. Note that the adsorption mode is sterically hindered for the complex suggested to be the precursor to S–(–)–methyl lactate as it is shown in figure 3B.

This molecular modelling approach leads to a reasonable explanation for the enantio–differentiation of the system, despite the fact that our theoretical prediction has been made for an ideal case, as the interaction with Pt surface atoms could not (and cannot yet) be considered. The calculations showed that – in agreement with the experimental observations – a change in the chirality of the stereogenic region (C_8, C_9) of the cinchona alkaloid used as modifier results in a corresponding change in the chirality of the product lactate. In the transition complex the substrate is bound to the modifier via N–H–O interaction of the protonated quinuclidine nitrogen and the oxygen of the α–carbonyl of pyruvate. This complex resembles a half–hydrogenated state of pyruvate and has already been suggested to be a probable transition state in the enantioselective hydrogenation of α–ketoesters [3].

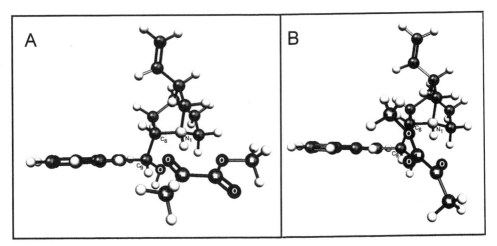

Figure 3. Our model : side view of the energetically most favourable complexes formed between protonated cinchonidine and methyl pyruvate which would yield R–(+)–methyl lactate (A) and S–(–)–methyl lactate (B), respectively, on hydrogenation. The oxygen atoms are marked with a white O and the carbon atoms of the reactant with a white square.

DISCUSSION

Comparison of the Models

A comparison of the various mechanistic models in the light of the experimental findings summarized in the introduction, reveals that none of the models can describe all experimental observations consistently. Assuming that the experimental findings are free from uncertainties, this clearly indicates that the models suggested so far do not describe the "real" mechanism sufficiently well. Experimental uncertainties can, however, not be ruled out due to several factors, which may disguise the experimental findings, such as the presence of impurities or trace amount of water [12], transformation of the alkaloid under reaction conditions [11, 13], catalyst deactivation due to (polymeric) by–product formation or any uncontrolled interaction between substrate, solvent and modifier before and during the hydrogenation reaction [23]. These circumstances make an absolute judgement of the models rather difficult.

All three models provide an explanation for enantio–differentiation. Although based on different concepts (1:1 interaction or ordered array), our model suggests a similar mechanism for the enantio–differentiation, i. e. R–(+)–lactate formation in the presence of cinchonidine, as does the template model [3,11,21]. We propose that a hydrogen bond exists between a protonated cinchonidine molecule and the α–carbonyl group of pyruvate ester and Wells *et al.* suggest that an important step of the mechanism is hydrogen bond formation between the half–hydrogenated state of pyruvate and quinuclidine nitrogen. Note that the enantio–differentiation to S–product formation is completely different in the two models. The "template model" requires a second, carefully positioned cinchonidine molecule to hinder adsorption of the pyruvate molecule in an orientation to give S–product, whereas in our calculations the 1:1 complex, which gives S–product, cannot easily adsorb on a flat Pt surface (although would at edge sites). Analysis of kinetic data for low concentrations of added modifier in the reaction system suggests strongly that a 1:1 relationship exists between cinchona alkaloid and pyruvate, in conflict to the requirements of the template model [19,24].

The only agreement between the latter two mechanisms and that of Augustine *et al.* [12] is in the importance of the quinuclidine nitrogen. Planar adsorption of the alkaloid through the quinoline ring system results in S–product formation, according to the model of Augustine *et al.*. The alkaloid is also in a different conformation to that found in the crystal structure, which is used by Wells' group and ourselves. The alkaloid conformation appears to have a potential energy above the calculated minimum [25]. Although interaction of the alkaloid with the Pt surface would be expected to modify the conformation of the molecule, the driving force to overcome the energy barriers is not clear in Augustine's model. Indeed, adsorption with the quinoline perpendicular to the surface would favour the rotation of the quinuclidine ring with N_1 orientated towards the surface, which is one of the calculable minimum energy conformations.

Augustine's model (figure 2D) [12] relies on N_1 being adjacent to the hydroxyl group (in N_1–C_8–C_9–O position) to enable "six–membered ring interaction" between the quinuclidine N_1 and the keto carbon atom as well as the C_9 oxygen and the ester carbon atoms. Note that high optical yields have been reported in many instances by applying modifiers with a basic nitrogen in a chiral environment and an oxygen atom in a 1,4 relationship [2]. The involvement of oxygen at the C_9 of cinchonidine into the transition complex is a real development towards interpreting enantio–differentiation. However, the assumption of a rigid six–membered ring suggests the same importance to N_1 and to oxygen at C_9 in the interaction between pyruvate and cinchonidine, which is against the experimental observations [13].

Another questionable point of this model is the key role of coordinatively unsaturated Pt atoms (corner, edge and adatoms) in enantio–differentiation. If, as Augustine *et al.* [12] propose, the presence of a Pt adatom is important for enantioselectivity in favour of R–product, it would be expected that selectivity would decrease with increasing temperature of catalyst pre–treatment, due to a decrease in the number of coordinatively unsaturated sites (annealing). In contrast, heat pre–treatment of the catalyst was reported to increase the optical yield [6–8]. The requirement of Pt particles above 3 nm to achieve high enantioselectivity [10,15] is also in contrast to what could be expected from the model of Augustine *et al.*. Note that the unusual reaction conditions (solvent and pressure), substantially different from those applied earlier, renders the comparison ambiguous.

The proposals by Augustine's and Wells' groups are quite specific in the location and positioning of alkaloid molecules for an asymmetric hydrogenation site. In the Augustine model adsorption of cinchonidine to produce S–(–)–product must occur adjacent to a corner atom. For the production of R–(+)–lactate the cinchonidine must adsorb adjacent to a Pt adatom. The template model positions the aromatic rings of the quinoline over adjacent Pt atoms (figure 2). The same is also true for the carbonyl groups of the pyruvate molecule. A second alkaloid molecule is required in a specific location to hinder S–(–)–lactate formation for a site in the centre of a Pt crystallite. At edge sites the requirement is that there is no room for the pyruvate molecule to adsorb in a manner to produce S–(–)–product on hydrogenation. Although evidence for the adsorption of molecules iso–electronic with the pyruvate exists [26], no evidence, (with the exception of a tenuous link to ordered adsorption of benzene co–adsorbed with CO on a Pt single crystal [27]) is known to suggest that the alkaloid molecules should adsorb as required.

Another argument against the specific location of cinchonidine is the co–adsorption of solvent on platinum. Wells and coworkers [11] already observed that ethanol, which is the most widely used solvent in pyruvate hydrogenation, adsorbs dissociatively on Pt:

$$CH_3CH_2OH \rightarrow CH_3CH_2O_{ad} + H_{ad} \tag{1}$$

However, the adsorption of primary alcohols on Pt does not stop at this stage and the final "product" of this irreversible, destructive adsorption process is linear and bridge-bonded CO, with some C–H type species via an aldehyde–like intermediate [28,29]:

$$CH_3CH_2O_{ad} \rightarrow CH_3CO_{ad} \rightarrow C_xH_{y,ad} + CO_{ad} \tag{2}$$

There is spectroscopic evidence [30] that CO coverage of Pt may be quite high after contacting with primary alcohols ($\Theta = 0.50$ in the case of ethanol). Consequently, during pre–modification of the catalyst, the cinchona alkaloid has to compete with CO and C–H type species on the Pt surface which reduces the likelihood of any ordered array formation supposed to occur on an ideally clean surface.

No explicit positioning of adduct, relative to individual atoms is made in our model. We supposed a planar adsorption of the quinoline ring in the protonated cinchonidine – pyruvate adduct on a flat Pt surface as a first approximation. This geometry seems to be the most probable according to the necessity of relatively big, thermally annealed Pt particles for achieving high selectivity [6,10,15]. We are currently investigating the adsorption of cinchonidine on Pt to supply further information to the modelling.

Our Model : Limitations and Future Development

Our theoretical model in its present state indicates that the outstanding selectivity measured in acetic acid [17] or after an acidic treatment of the catalyst [8] is –at least partly– due to the protonation of the alkaloid. NMR, UV and chromatographic studies proved that cinchonidine is in a protonated state in acetic acid and, astonishingly, in ethanol after the usually applied aerobic pretreatment of the catalyst slurry containing the modifier [31]. We are currently working on a more sophisticated model (i) to involve the interaction of pyruvate with the lone electron pairs of oxygen at C_9, which seems to be necessary for achieving high enantioselectivity [13], and (ii) on the extension of calculations of the interaction between cinchonidine and pyruvate ester in neutral aprotic solvents such as toluene.

It is clear from the above discussion that the most dubious part of the mechanistic models is the nature of cinchonidine and pyruvate adsorption on Pt. We can discriminate between two functional parts of the modifier, the quinoline ring, which anchors the modifier on the platinum surface and the quinuclidine entity which is responsible for the enantio-differentiation. Wells and coworkers and ourselves assumed flat adsorption of the quinoline ring of cinchonidine, parallel to the Pt surface. We consider it as a first and reasonable approximation, which should be refined later when enough experimental evidence will be available.

Augustine *et al.* [12] proposed that there are at least two different ways for cinchonidine to adsorb on Pt, each having different enantioselectivity. Though there is no direct evidence yet, it is feasible that the orientation of the alkaloid and the number of Pt atoms occupied by a single alkaloid molecule (or more general by the transition complex) is a function of the alkaloid:Pt ratio. It is well known that the mode of CO adsorption (linear or bridged) depends largely on the coverage of the metal surface by this species [32]. A closer analogy is the study of the adsorption of quinoline and other O– and N–containing organic molecules on various metals. The adsorption of diphenols and chinons on Pt from aqueous solutions produced flat–oriented (π–bonded) species at concentrations below 10^{-4} M and edge–oriented (di–σ–bonded) species at above 10^{-3} M [33]. Similarly, two orientations of α–quinoline and three orientations of β–quinoline were observed on mercury as a function of concentration and electrode potential [34]. A change in electrode potential may formally be interpreted as a change in hydrogen pressure according to the Nernst equation, which would offer a possibility for the explanation of the pressure–dependence of enantioselectivity [9]. The assumption of at least two different modes of adsorption of cinchonidine would also permit the interpretation of the loss of enantio–differentiation observed above room temperature [9]. The concentration dependence of alkaloid adsorption is the likely explanation to the early observation by Orito et al. [5] that in methyl benzoylformate hydrogenation there is an optimum in cinchonidine concentration above which enantioselectivity decreases markedly.

It was reported recently [35] that the modifier may be anchored to the support via the quinoline ring and the interaction of the quinuclidine part with the substrate on Pt is sufficient for obtaining enantioselectivity comparable to selectivity obtained with the alkaloid adsorbed on Pt. This would mean that the adsorption of cinchona alkaloid on Pt is not a necessary pre–requisite of enantio–differentiation. However, the authors have yet to prove that the observed good selectivity is not due to the migration of formerly immobilized modifier onto the Pt surface.

If we accept that there are at least two different modes of adsorption of cinchonidine (and of the transition complex) on Pt, we can think on the possibility of the existence of more than one single mechanism for enantioselective hydrogenation of α–ketoesters. Thus, specific information of the adsorption of the modifier and the reactant are crucial to find an unambiguous mechanism.

CONCLUSIONS

The three mechanistic models suggested in the past five years for the interpretation of enantio–differentiation in the hydrogenation of α–ketoesters over cinchona–modified Pt catalysts, contain reasonable elements which explain a part of the catalytic results. However, none of them offers a comprehensive explanation for the most important experimental observations. Substantial further research concerning the interaction between substrate, solvent, modifier and catalyst surface is necessary for improvement of the models and to minimize the necessity for speculation.

ACKNOWLEDGEMENT

Financial support by the Swiss National Science Foundation is kindly acknowledged.

REFERENCES

[1]   H. U. Blaser, Tetrahedron: Asym., 2 (1991) 843.
[2]   H. U. Blaser, Chem. Rev., 92 (1992) 935.
[3]   G. Webb and P. B. Wells, Catal. Today, 12 (1992) 319.
[4]   A. Baiker, in Proc. DGMK Conference on "Selective Hydrogenation and Dehydrogenation", Kassel, Germany, 1993, p. 119.
[5]   Y. Orito, S. Imai, S. Niwa and N. G. Hung, J. Synth. Org. Chem. Jpn., 37 (1979) 173.
[6]   Y. Orito, S. Imai and S. Niwa, J. Chem. Soc. Jpn., (1979) 1118.
[7]   Y. Orito, S. Imai and S. Niwa, J. Chem. Soc. Jpn., (1980) 670.
[8]   S. Niwa, S. Imai and Y. Orito, J. Chem. Soc. Jpn., (1982) 137.
[9]   H. U. Blaser, H. P. Jalett, D. M. Monti, J. F. Reber and J. T. Wehrli, in M. Guisnet et al. (Eds.), Stud. Surf. Sci. Catal., Vol. 41, Elsevier, 1988, p. 153.
[10] J. T. Wehrli, A. Baiker, D.M. Monti and H. U. Blaser, J. Mol. Catal., 49 (1989) 195.
[11] I. M. Sutherland, A. Ibbotson, R. B. Moyes and P. B. Wells, J. Catal., 125 (1990) 77.
[12] R. L. Augustine, S. K. Tanielyan and L. K. Doyle, Tetrahedron: Asymm., 4 (1993) 1803.
[13] H. U. Blaser, H. P. Jalett, D. M. Monti, A. Baiker and J. T. Wehrli, in R. K. Grasselli and A. W. Sleight (Eds.), Stud. Surf. Sci. Catal., Vol. 67, Elsevier, 1991, p. 147.
[14] J. L. Margitfalvi, P. Marti, A. Baiker, L. Botz and O. Sticher, Catal. Lett., 6 (1990) 281.
[15] J. T. Wehrli, A. Baiker, D.M. Monti and H. U. Blaser, J. Mol. Catal., 61 (1990) 207.
[16] J.T. Wehrli, A. Baiker, D.M. Monti, H.U. Blaser and H.P. Jalett, J. Mol. Catal. 57 (1989) 245.
[17] H. U. Blaser, H. P. Jalett and J. Wiehl, J. Mol. Catal., 68 (1991) 215.
[18] J. T. Wehrli, Thesis, ETH, Zürich, 1989.
[19] M. Garland and H. U. Blaser, J. Am. Chem. Soc., 112 (1990) 7048.
[20] G. Bond, P. A. Meheux, A. Ibbotson and P. B. Wells, Catal. Today, 10 (1991) 371.
[21] O. Schwalm, B. Minder, J. Weber and A. Baiker, Catal. Lett., 23 (1994) 271.
[22] O. Schwalm, J. Weber, J. Margitfalvi and A. Baiker, J. Mol. Structure, 297 (1993) 285.
[23] J. L. Margitfalvi, B. Minder, E. Talas, L. Botz and A. Baiker, in Guczi et al. (Eds.), Stud. Surf. Sci. Catal., Vol. 75, Elsevier, 1993, p. 2471.
[24] K. E. Simons, A. Ibbotson and P. B. Wells, in T. J. Dines, C. H. Rochester and J. Thomson (Eds.), Catalysis and Surface Characterisation, R. S. C., 1992, p. 174.
[25] K. E. Simons, A. Ibbotson and P. B. Wells, unpublished results.
[26] A. T. Bates, Z. K. Leszczynski, T. T. Phillipson, P. B. Wells and G. R. Wilson, J. Chem. Soc. A., (1970) 507.
[27] C. M. Mate and G. A. Somorjai, Surf. Sci., 160 (1985) 542.
[28] M. W. Breiter, in J. O´M. Bockris and B. E. Conway (Eds.), Modern Aspects of Electrochem., Vol. 10, Plenum Press, 1975, p. 161.
[29] R. Parsons and T. VanderNoot, J. Electroanal. Chem., 257 (1988) 9.
[30] L. W. H. Leung and M. J. Weaver, Langmuir, 6 (1990) 323.
[31] T. Mallat, B. Minder, P. Skrabal and A. Baiker, Catal. Lett., (submitted).
[32] T. A. Dorling and R. L. Moss, J. Catal., 7 (1967) 378.
[33] M. P. Soriaga, E. Binamira–Soriaga, A. T. Hubbard, J. B. Benzinger and K. W. P. Pang, Inorg. Chem., 24 (1985) 65.
[34] M. W. Huphreys and R. Parsons, J. Electroanal. Chem., 82 (1977) 369.
[35] H. U. Blaser and M. Müller, 1st European Conf. on Catal., Book of Abstracts, 1993, p. 406.

STUDIES OF THE PLATINUM CINCHONA ALKALOID CATALYST FOR ENANTIOSELECTIVE α-KETOESTER HYDROGENATION

P. J. Collier, T. Goulding, J. A. Iggo and R. Whyman

Liverpool University
Department of Chemistry
P.O. Box 147
Liverpool, L69 3BX, United Kingdom

INTRODUCTION

The enantioselective hydrogenation of α–ketoesters catalysed by a cinchona alkaloid modified supported platinum catalyst was first reported by Orito in 1978 [1]. Since then extensive research by Wells and co–workers and Blaser, Baiker and co–workers has established the optimal reaction conditions and suggested a reaction mechanism [2,3]. Modification of the platinum surface by naturally occurring cinchona alkaloids not only effects the product enantioselectivity, cinchonidine gives the (R)–product whilst cinchonine, its pseudo enantiomer, furnishes the (S)–product, (schemes 1 and 2), but it also enhances the reaction rate by up to one hundred times over the racemic reaction. The reaction is well behaved giving quite respectable optical yields, as high as 94% under optimised conditions. However the many variables associated with the physical and chemical properties of supported catalysts make mechanistic elucidation difficult. In order to alleviate these problems we have been examining the application of novel colloidal catalysts to this reaction. Colloids have been shown to be more active than analogous conventional heterogeneous catalysts because of an increased active metal surface area [4]. This feature may allow the enantioselective hydrogenation of less reactive substrates than α–ketoesters, thus expanding the presently limited scope of the reaction. Colloidal metals can be prepared cleanly by condensing metal vapour into an organic solvent at 77 K (metal vapour synthesis), leaving none of the surface halide contaminants that often reside on supported catalysts, contaminants which can interfere with the adsorption characteristics of the surface and modifier. To a certain extent colloidal particle size and morphology can be influenced by preparation procedure, for example different particle types may be produced by the use of different solvents, allowing catalysts to be tailored to particular needs. For these reasons colloidal metals make attractive catalyst models, free from complicating support effects. We ultimately hope that our studies will lead to improved catalysts for asymmetric hydrogenation, able to deal with a wider range of substrates, and also that a greater understanding of the reaction mechanism might be gained.

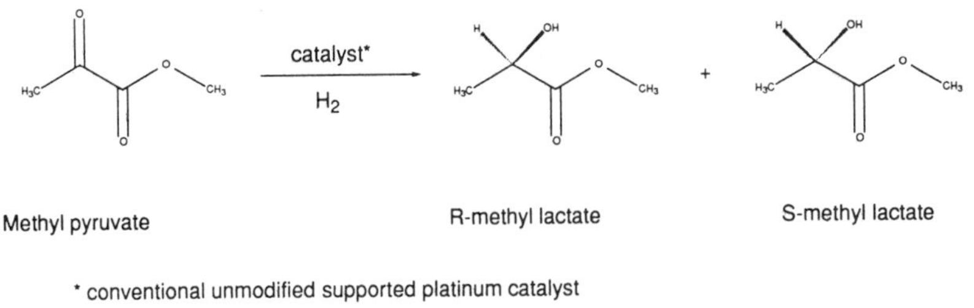

Methyl pyruvate   R-methyl lactate   S-methyl lactate

* conventional unmodified supported platinum catalyst

Scheme 1.

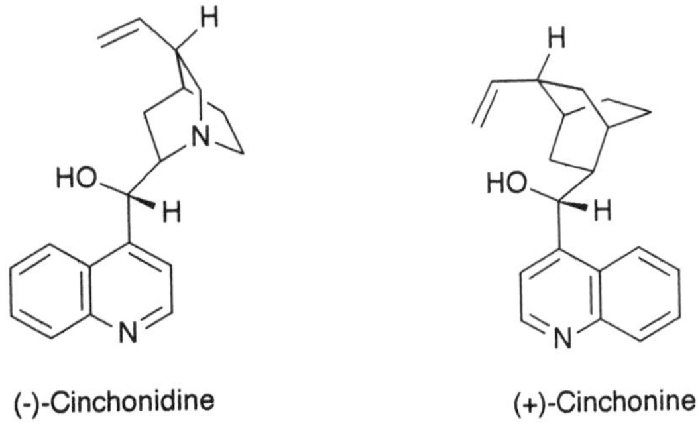

(-)-Cinchonidine   (+)-Cinchonine

Scheme 2.

EXPERIMENTAL

Solutions of the platinum colloids were prepared by metal vapour synthesis techniques using a commercial positive hearth electrostatically focused electron beam rotary metal atom reactor (Torrovap Industries, Inc., Ontario, Canada). Typically 0.1 g platinum was evaporated over 3 h and co–condensed at 77 K with *ca.* 200 ml degassed, dried 2–butanone. After slow warm–up and melting the resultant brown liquid was transferred under anaerobic conditions from the reactor flask into a Schlenk receiver vessel. In one synthesis KD1, a proprietary dispersing agent manufactured by ICI, was added (at *ca.* 0.1 wt% concentration in 2–butanone) under vacuum into the rotary reactor flask as the metal/organic matrix began to soften during the warm–up procedure (Colloid 1).

Phase contrast images of the colloidal particles were obtained with a top entry Philips EM400 electron microscope operating at an accelerating voltage of 120 kV. A lacey carbon film on a copper grid (2.8 mm) support was used.

The solvent requirements of the asymmetric hydrogenation reaction and those of colloid preparation must be reconciled, the former preferring a non polar medium, and the latter a relatively polar coordinating solvent. 2–Butanone was chosen since it has a fairly low dielectric constant ($\varepsilon = 18$) with good coordinating properties for colloidal stabilization. In some reactions ethanol was added to improve the solubility of the alkaloid.

The "conventional" heterogeneous catalysts, EUROPT–1 (Johnson Matthey, 6.3% Pt/SiO$_2$) and ESCAT 24 (Engelhard, 5% Pt/Al$_2$O$_3$) were reduced in hydrogen prior to use. Methyl pyruvate (Fluka) and cinchonidine (Aldrich) were used as received.

Three catalyst modification procedures have been described for methyl pyruvate asymmetric hydrogenation [5, 6], and have been adapted for these studies as follows:

(A) Catalyst stirred under nitrogen atmosphere in a 2–butanone:ethanol (1:1) cinchonidine solution (0.022 mol dm^{-3}) overnight (referred to as "anaerobic modification") and then loaded into an adapted heavy walled glass Schlenk tube containing methyl pyruvate (0.1 mol) with air carefully excluded throughout these steps. Reaction proceeded under 5 bar hydrogen pressure with stirring at 398 K for 8 hours (referred to as "anaerobic modification") [5].

(B) As above except the catalyst/cinchonidine solution was stirred in air (referred to as "aerobic modification") [5].

(C) Catalyst loaded into a mechanically stirred, high pressure autoclave containing cinchinodine, methyl pyruvate (0.2 mol) and 2–butanone, carefully excluding air (referred to as "in situ modification") [6].

All conversions were determined using a gas–chromatograph (FFAP column DANi OV1) and calibrated against cyclohexanone as internal standard. Optical yields were determined by polarimetry.

RESULTS

Table 1.

Experiment number	Catalyst	Surfactant	Reaction procedure	Conversion (%)	Optical yield (%)
1	Colloid 1	✓	A	24	10
2	Colloid 1	✓	B	100	12
3	Colloid 1	✓	B[a]	7	2
4	Colloid 2	x	B	100	16
5	Colloid 2	x	B[b]	27	16
6	Colloid 3	x	B	100	13
7	EUROPT–1		A	4	n/r
8	EUROPT–1		B	100	30
9	EUROPT–1		B[c]	20	31
10	ESCAT 24		C	3	n/r
11	Colloid 4	x	C	17	40

[a] As for procedure B, except reaction temperature 298 K rather than 308 K.
[b] When the catalyst had been modified it was subjected to ultrasound treatment for 15 minutes, otherwise procedure B followed as normal.
[c] As for procedure B, except reaction performed in the high shear mechanically stirred autoclave used in procedure C, with ethanol only as solvent, and reaction time 18 hours, hydrogen pressure 14 bar, temperature 298 K.

Characterisation of Colloidal Particles

Preparations of the colloidal samples, of concentrations ranging between 1 and 5 mg Pt.ml^{-1} were, after evaporation onto copper grids, imaged by high resolution transmission electron microscopy. Micrographs of colloids 2, 3 and 4 (see Table 1) were basically similar and showed the presence of platinum "super–spheres" of *ca.* 200 nm in diameter which, under higher magnification (410K), were revealed to comprise of agglomerations of very small primary particles of *ca.* 2 nm diameter. Electron micrographs of colloid 1, prepared similarly from Pt/2–butanone but in the presence of the dispersing agent KD1, confirmed the presence of the primary particles with a size distribution ranging from 1 – 3 nm. Figures 1 and 2 show colloid 2 before and after modification by procedure B.

Micrographs of the colloid after the full reaction procedure showed further agglomeration, but the basic particles remained intact.

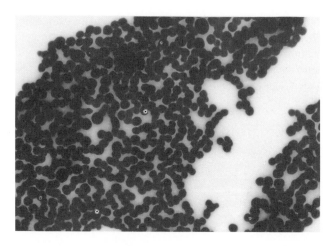

Figure 1. Electron micrograph of colloid 3, taken shortly after preparation. Magnification 32K. See main text for further details.

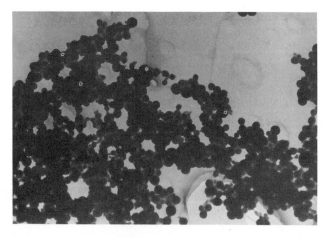

Figure 2. Electron micrograph of colloid 2 after modification, prior to the asymmetric hydrogenation reaction (see table 1 for reaction conditions). Magnification 32K. For further details see main text.

Catalytic performance of colloids

Table 1 compares the conversions and selectivities achieved in methyl pyruvate hydrogenation using colloidal and conventional supported metal catalysts under comparable, but non–optimised conditions. In all cases the activities and selectivities are lower than those reported by Wells and Blaser under optimised conditions [5,6], although efforts were made to follow the procedures described by these groups.

For colloidal catalysts 1, 2 and 3 the aerobic procedure described by Wells (our procedure B) (runs 2,4,6) gave higher conversions and selectivities than the anaerobic procedure A (run 1) as expected on the basis of Wells' results. The conversions achieved using the colloidal catalysts exceeded those obtained for the conventional catalyst, EUROPT–1 (runs 7 and 8). The best selectivity of all runs was observed following Blaser and Baiker's anaerobic procedure (C) in which the catalyst modification is carried out *in situ* (run 11). Again the conversion achieved with the colloidal catalyst exceeded that of the conventional catalyst, ESCAT 24 (run 10). The low conversion in run 10 precluded measurement of the optical yield in this reaction.

The importance of the "super–spheres" observed by electron microscopy was examined by treating a sample of colloid 2 with ultrasound to disperse the super–spheres after the modification step. Catalyst activity is greatly reduced though selectivity remains unchanged (compare entries 4 and 5, Table 1). However, in contrast to this observation the addition of KD–1, the surfactant that inhibits the formation of the super–spheres, to colloid 1 did not appear to affect the activity of the catalyst system. (compare entries 2 and 4)

DISCUSSION

Orito's platinum/cinchona alkaloid catalyst for the enantioselective hydrogenation of α –ketoesters is one of two heterogeneous catalysts capable of high enantiomeric excesses in asymmetric catalysis. The origin of the enantioselectivity has therefore been of prime interest [5], followed by attempts to optimise the reaction [5,6] and extend its application to more useful substrates [7].

This preliminary study has addressed the first and last of these points. It is encouraging to note that, without optimisation of the reaction conditions, the conversions achieved by the colloidal catalysts in every case equal or exceed those we have found for the conventional heterogeneous catalysts under comparable conditions. If this activity can be optimised and maintained it gives hope for extending the reaction to less activated substrates.

The effect of colloid preparation and modification with cinchonidine in combination with other reaction variables has begun to be addressed. The best optical yields are obtained in 2–butanone using *in situ* modification as described by Baiker and Blaser, rather than the aerobic modification as described by Wells, suggesting that surface oxygen is not necessary to promote the selectivity of the reaction using the colloidal catalysts. The colloidal particle size and morphology can be controlled by the addition of a surfactant: surfactants favouring smaller particle sizes and inhibiting the formation of the superspheres. It is interesting to note that the sample containing surfactant (colloid 1) gave lower optical yields. It has been suggested that particle sizes above 2 nm are necessary for high enantioselectivity [2] although it is not possible at this time to attribute the low enantioselectivity of colloid 1 definitively to a particle size effect rather than an interference in the surface reaction by the surfactant.

CONCLUSIONS

These preliminary results indicate that, under comparable non–optimised reaction conditions, the best activities and selectivities that were achieved with colloidal catalysts matched those of conventional catalysts. We shall expand our studies with these novel and interesting materials in an effort to improve their selectivity in the asymmetric hydrogenation of α–ketoesters.

ACKNOWLEDGEMENTS

We thank SERC for financial support, P.B.Wells for helpful discussions, R. Parry (Engelhard U.K) for supplying the ESCAT 24 and Prof. P.B. Wells and Prof G. Webb for samples of EUROPT–1.

REFERENCES

[1]   Y. Orito, S. Imai and S. Niwa, "Collected papers of the 43rd Catalyst Forum", Japan, 1978, p.30.
[2]   G. Webb and P. B. Wells, *Catalysis Today*, **12**(1992), 319.
[3]   H. U. Blaser, *Tetrahedron: Asymmetry* **2** (1991), 843.
[4]   J.S.Bradley, E.Hill, M.E.Leonowicz and H.Witzke, *J. Mol. Catal.*, **41** (1987), 59.
[5]   Wells modification procedure.
        (a) I. M. Sutherland, A. Ibbotson, R. B. Moyes and P. B. Wells, *J. Catal.*, **125** (1990) 77.
        (b) P. Meheux, A. Ibbotson and P. B. Wells, *J. Catal.*, **128** (1991), 387.
[6]   Baiker and Blaser modification procedure.
        (a) J. T. Wehrli, A. Baiker, D. M. Monti and H. U. Blaser, *J. Mol. Catal.*, **49** (1989), 195.
        (b) J. T. Wehrli, A. Baiker, D. M. Monti, H. U. Blaser and H. P. Jalett, *J. Mol. Catal.*, **57** (1989), 245.
        (c) H. U. Blaser, H. P. Jalett and J. Wiehl, *J. Mol. Catal.*, **68** (1991), 215.
[7]   W.A.H. Vermeer, A.Fulford, P.Johnston and P.B.Wells, *J. Chem. Soc., Chem. Commun.*, (1993), 1053.

ASYMMETRIC HYDROGENATION OF ETHYL PYRUVATE USING Pt–CONTAINING ZEOLITES MODIFIED WITH (–)–CINCHONIDINE

W. Reschetilowski[1], U. Böhmer[1], J. Wiehl[2]

[1]Karl–Winnacker–Institut der DECHEMA e.V.
   Theodor–Heuss–Allee 25,
   D–60486 Frankfurt am Main, Germany
[2]W.C. Heraeus GmbH
   Heraeusstraße 12–14
   D–63450 Hanau

ABSTRACT

Platinum loaded, (–)–cinchonidine modified zeolites of the types Y, ZSM–35, β and ZSM–5 manifest high catalytic activity for the asymmetric hydrogenation of ethyl pyruvate to ethyl lactate. The catalysts based on Y and ZSM–35 zeolites show the highest enantio-selectivity. To obtain the highest reaction rates and the optical yields the Pt content should be around 5 wt%

Enantiomeric excess is strongly influenced by the type of solvent used and can be raised from 60% ee. to 80% ee. merely by varying the solvents.

INTRODUCTION

Besides the biochemical production of enantiomerically pure compounds, non–biochemical asymmetric syntheses in the presence of heterogeneous catalyst systems are fast gaining ground.

The first heterogeneously catalysed enantioselective reactions were reported by Izumi et al. [1] in 1956. They hydrogenated oxime and oxazolone derivates by means of a silk Pd catalyst. From the beginning of the sixties until now, various research groups have carried out numerous investigations on the use of modified Raney nickel as catalyst [2,3] in the production of divers chiral compounds.

In 1979 Orito et al. [4] described the enantioselective conversion of methyl benzoylformate to R–(–)–methyl mandelate in the presence of Pt–containing carbon. These studies were fundamental to further research by Blaser, Baiker and others [5–7] on enantioselective hydrogenation of various α–keto esters to α–hydroxy esters with Pt containing SiO_2 or Al_2O_3 as catalysts .

To date, however, only very few publications have dealt with the potential of using modified zeolites as catalysts for enantioselective reactions [8–10].

The aim of this study is to demonstrate the aptitude of Pt–containing zeolites of varying structure for enantioselective catalytic hydrogenation from ethyl pyruvate to R–(+)–ethyl lactate (figure 1).

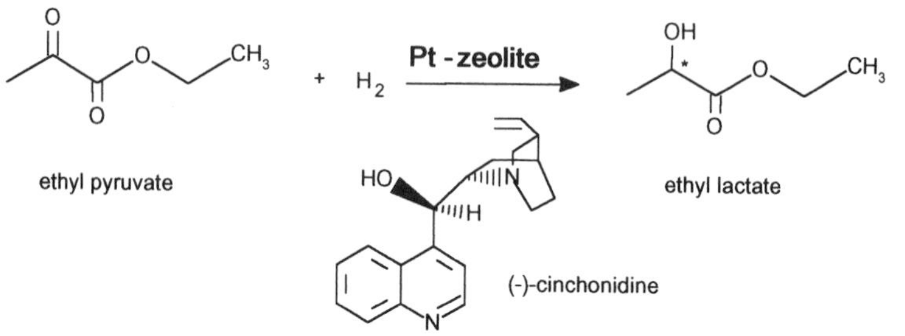

Figure 1. Reaction scheme of the hydrogenation from ethyl pyruvate to ethyl lactate.

EXPERIMENTAL

Materials

The carriers for the active hydrogenation components used in these investigations were zeolites Y, ZSM–5, ZSM–35 and β in the sodium and the hydrogen forms. Table 1 provides several important specifications of the types of zeolite used.

Table 1. Specifications of the zeolites used.

	Aluminium-rich		Silicon-rich	
type of zeolite	Y	ZSM-35	β	ZSM-5
Si/Al ratio	2.5	5	12.5	15
pore size [nm]	12-ring 0.71	10-ring 0.42·0.54 8-ring 0.35·0.48	12-ring 0.64·0.76 12-ring 0.55·0.55	10-ring 0.53·0.56 10-ring 0.51·0.55
particle size [µm]	2.9	9.6	12.0	8.7

Platinum was used as the active hydrogenation compound. The platinum content of the zeolites was varied between 1 and 5 wt%. The impregnation procedure is described in detail in reference [11].

Methods

Ethyl pyruvate was hydrogenated in a 250 ml laboratory autoclave at 20 °C with initial hydrogen pressure of 70 bar under constant stirring at 1200 min^{-1}. In all tests the liquid/solid ratio was kept constant with 10 ml ethyl pyruvate, 20 ml solvent, and 40 mg (−)−cin−chonidine as the chiral auxiliary continually added to 200 mg of the catalyst.

The conversion of ethyl pyruvate was monitored by measuring the pressure drop in relation to time. The calculations of the initial reaction rate r_0 were based on this drop in pressure. Conversion and enantiomeric excess were determined by using GC MS (Varian GC, ITD−800), equipped with a 30 m capillary column with permethyl−β−cyclo−dextrin/polysiloxane as the stationary phase (C−Dex−B^{TM}, J&W). Analytic conditions can be found in [11].

RESULTS AND DISCUSSION

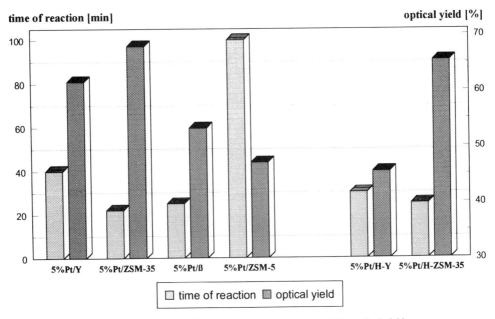

Figure 2. Effect of zeolites used on the reaction time and the optical yield.

Initially comparative investigations were carried out under the given conditions to establish the suitability of the prepared Pt−containing zeolites for hydrogenating ethyl pyruvate to ethyl lactate. In the first instance methanol was used as the solvent. Figure 2 shows a comparison of obtained enantiomeric excesses and the necessary reaction times for the catalysts used. These values vary according to the type and composition of the zeolitic carrier material. As can be seen in figure 2, the silicon−rich, medium−pored ZSM−5 produces a significantly lower reaction rate of hydrogenation than the other three zeolites (Y, ZSM−35, β) in the sodium form. The highest reaction rates are to be observed both for the silicon−rich, fairly wide−pored zeolite β and for the aluminium−rich, relatively narrow−pored zeolite ZSM−35. This is an indication that the accelerating effect during hy−

drogenation of ethyl pyruvate is not due to pore geometry and pore size or to the lattice composition of the zeolite. This assertion is corroborated by the reaction rate obtained for the aluminium–rich, wide–pored zeolite Y which had a mean value in the series of investigated zeolites in the sodium form. If the aluminium–rich Y and ZSM–35 zeolites are converted into the hydrogen form by cation exchange with an aqueous 0.5 n NH_4NO_3 solution and subsequent thermal deammonisation, and are then used in this form for the hydrogenation reaction, scarcely any difference in the reaction time can be observed in the case of H–ZSM–35 and only a slight rise in that of zeolite H–Y (cf. figure 2). The fact that a low reaction time and the ensuing high reaction rate do not encourage any prediction of attainable enantiomeric excess is also evident from figure 2. The two aluminium–rich zeolites Y and ZSM–35, with 62% and 67% ee. respectively, permit significantly higher optical yields, compared with 53% and 47% ee. for the silicon–rich zeolites β and ZSM–5 respectively. When using zeolite H–Y, as opposed to zeolite Y in the sodium form, a decrease in the reaction time contrasts strongly with the obvious deterioration in selectivity (from 62% to 46% ee.) However, compared with the sodium form, H–ZSM–35 evinces almost no change in catalytic properties. As no improvement in selectivity could be achieved by using zeolitic carriers in the hydrogen form future, investigations will be confined to sodium forms.

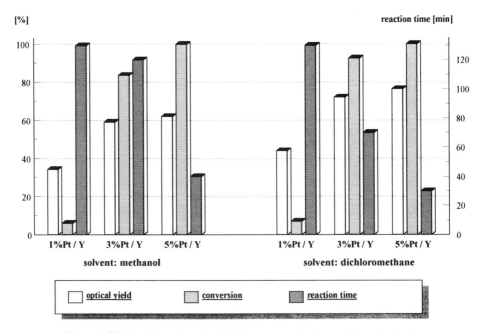

Figure 3. Effect on Pt content of zeolite Y on the reaction time and the optical yield.

First investigations to reduce the Pt content of the catalysts have been carried out. The carrier for this study was the zeolite Y in the sodium form. To obtain information on the influence of the solvent on this hydrogenation, methanol and dichloromethane were used. As can be seen in figure 3, it is possible to get almost the same enantioselectivity with catalyst 3% Pt/Y as with 5% Pt/Y. However, it takes 70 min to get a conversion of 92% when dichloromethane is used as the solvent. This is twice as long as for the same reaction with 5% Pt/Y. If the Pt content is reduced to 1 wt% the results are very poor due to the very long reaction time (130 min), an optical yield of only 42% ee. and a conversion of 7%. The conclusion drawn from these facts is that the minimal Pt content for an effective hydrogenation of ethyl pyruvate must be about 5 wt%. In the case of dichloromethane as the solvent, an increase in the optical yield can be observed.

In order to examine the influence of the solvent on the enantioselectivity, catalyst systems based on the aluminium–rich Y and ZSM–35 zeolites were selected as they give the highest optical yields in the hydrogenation of ethyl pyruvate to ethyl lactate. Figure 4 presents the concentration–time curves for catalysts based on zeolites Y and ZSM–35 with various solvents. In the case of zeolite Y no essential differences in the hydrogenation curves could be observed (figure 4a). Only with methanol could a lower reaction rate be found than in the syntheses with other solvents. The concentration–time curves yield the same results for zeolite ZSM–35 (figure 4b). One striking difference is the course of the curve when dichloromethane is the solvent. In this case we did not calculate r_0 because the reaction does not follow a rate law of zero order.

When comparing these zeolites, it is evident that the initial reaction rates for zeolite ZSM–35 are 1.5 to almost 3 times higher than with zeolite Y under the same conditions. Table 2 gives a comparison of the calculated initial rates for these two zeolites and a 5% Pt/Al$_2$O$_3$ [5] in the presence of selected solvents.

Table 2. Influence of solvent on the catalytic behaviour of the Pt–containing catalysts.

Solvent	Dielectric constant ε	5% Pt/Y		5% Pt/ZSM-35		5% Pt/γ-Al$_2$O$_3$[a]	
		r_0[b]	ee [%][c]	r_0	ee [%]	r_0	ee [%]
methanol	32.63	0.37	61.6	1.1	69.8	1.08	75.1
ethanol	24.3	0.42	67.7	1.2	64.5	2.28	75.0
acetone	20.7	0.57	69.8	1.17	73.3	—	—
2-propanol	20.1	0.88	64.3	2.01	71.3	3.05	76.7
2-butanol	15.5	0.84	68.7	1.69	72.3	—	—
dichloromethane	9.08	0.69	76.3	—	75.0	2.20	81.4
acetic acid	6.15	0.56	83.9	0.69	86.1	—	—
cyclohexane	2.02	0.80	74.5	1.79	77.5	2.94	79.9
n-hexane	1.89	0.74	71.3	2.07	74.7	—	—

[a] results from ref. 5   [b] activity r_0[mmol · s^{-1} · g$^{-1}_{catalyst}$]
[c] optical yield

The initial reaction rates are readily comparable with the values for optimized Pt–containing SiO$_2$ and Al$_2$O$_3$ catalysts given in the literature [5,6]. A glance at the optical yields obtained with catalysts based on zeolites Y and ZSM–35 shows that the enantioselectivity of both carriers depends on the type of solvent used.

It is apparent that the best results can be obtained with solvents whose dielectric constants are less than 10 (e.g. n–hexane, dichloromethane). This outcome tallies with the

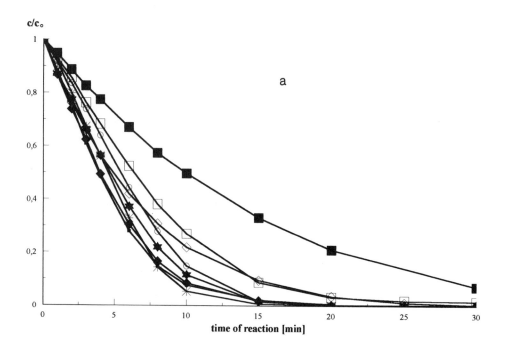

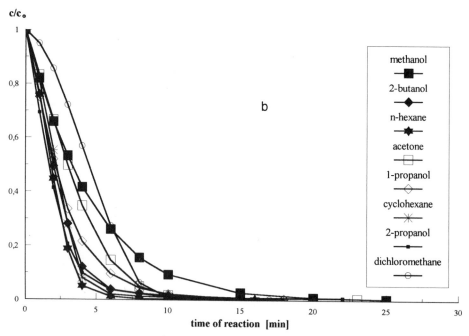

Figure 4. Kinetic curves for hydrogenation of ethyl pyruvate over 5% Pt/Y (a) and 5% Pt/ZSM-35 (b) with various solvents.

findings of Wehrli et al. [5] who also established that enantioselectivity is dependent upon the dielectric constant (figure 5).

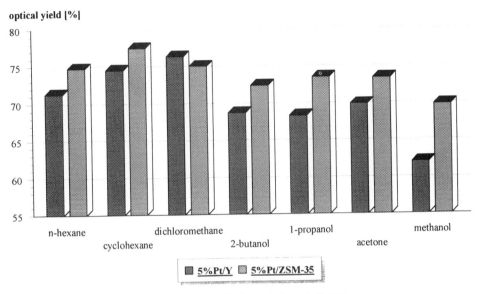

Figure 5. Influence of solvent on the optical yield.

Finally, in one experiment (+)–cinchonine was used as the auxiliary instead of (–)–cinchonidine. The result was the expected inversion of the optical yield (figure 6). This example shows that it is possible to get R–(+)–ethyl lactate or S–(–)–ethyl lactate just by varying the chiral auxiliary.

CONCLUSIONS

This current research is the first to demonstrate that Pt–containing zeolites are suitable catalyst systems for the enantiomeric conversion of ethyl pyruvate to ethyl lactate. First optimizations of this reaction have been achieved by varying zeolite structure, Pt content and solvent. Catalyst systems based on the aluminium–rich zeolites Y and ZSM–35 manifest the best enantioselectivity for this hydrogenation reaction.

It is intended to extend these investigations in order to optimize this reaction by varying the amounts of catalyst and chiral auxiliary used. Parallel tests will examine the application of available catalyst systems for enantioselective hydrogenation of other α–keto esters.

ACKNOWLEDGEMENT

The authors thank the AiF for support of this project which was funded by the Bundesministerium für Wirtschaft.

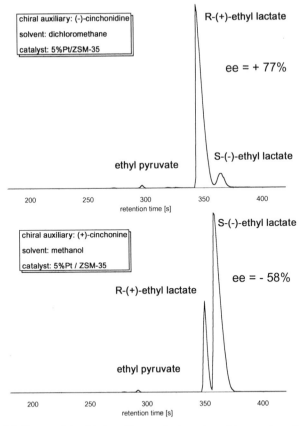

Figure 6. Influence of the chiral auxiliary on the inversion of the enantiomeric excess.

REFERENCES

[1]   S. Akabori, Y. Izumi, Y. Fujii and S. Sakurai, *J. Chem. Soc. Japan*, Pure Chem. Sec. **77** (1956) 1374.
[2]   I. Yasumori, M. Yokozeki and Y. Inoue, *Faraday Disc. Chem. Soc.* **72** (1981), 385.
[3]   T.Osawa, T. Harada and A. Tai, *J. Catal.* **121** (1990) 7.
[4]   Y. Orito, S. Imai, S. Niwa and Nguyen G.–H., *J. Synth. Org. Chem. Jpn.* **37** (1979) 173.
[5]   J.T. Wehrli, A. Baiker, D.M. Monti, H.U. Blaser and H.P. Jalett, *J. Mol. Catal.* **57** (1989) 245.
[6]   H.U. Blaser, H.P. Jalett and J. Wiehl, *J. Mol. Catal.* **68** (1991) 215.
[7]   J.L. Margitfalvi, B. Minder, E. Talás, L. Botz and A. Baiker, *Stud. Surf. Sci. Catal.* **75**C (1993) 2471.
[8]   R.M. Dessau, US Patent 4 554 262 (1985).
[9]   J. Weitkamp, *Stud. Surf. Sci. Catal.* **65** (1991) 21.
[10] A. Corma, A. Iglesias, C. del Pino, and F. Sánchez, *Stud. Surf. Sci. Catal.* **75**C (1993) 2293.
[11] W. Reschetilowski, U. Böhmer, Proc. DGMK–Conference "Selective Hydrogenations and
      Dehydrogenations", Kassel, Germany, 1993, p. 275.

SECTION III

HYDROGENATION SYSTEMS :

BROADENING THE SCOPE

HETEROGENEOUS CATALYTIC REACTIONS USING CHIRAL AUXILIARIES AND PRODUCING ENANTIOMERS IN EXCESS

A. Tungler[1], T. Tarnai[1], T. Máthé[2], J. Petró[1], R. A. Sheldon[3]

[1]Technical University of Budapest
  Department of Organic Chemical Technology
[2]Hungarian Academy of Sciences
  Research Group for Organic Chemical Technology
  Müegyetem rkp. 3
  H-1111 Budapest, Hungary
[3]University of Technology Delft
  Department of Organic Chemistry and Catalysis
  Julianalaan 136
  NL-2628 BL Delft, The Netherlands

INTRODUCTION

A perusal of the history of organic chemistry and catalysis reveals that more and more emphasis is being placed on asymmetric reactions. The great personalities in organic chemistry, such as Prelog and Corey, very early recognized the importance of asymmetric reactions in organic synthesis. In the last two decades many new methods have been discovered and applied.

The source of chirality in these reactions can be the substrate itself, an auxiliary, a catalyst or the solvent. From the viewpoint of both elegance and economic viability the source of chirality should be readily recyclable or, preferably, used in catalytic quantities.

In this review we have collected examples of reactions involving heterogeneous catalysts in conjunction with a chiral auxiliary or modifier. In many instances it is not easy to distinguish between the role of a chiral additive as an auxiliary (reaction with the substrate in solution) or as a modifier (reaction with the catalyst at the surface) : for example, in the hydrogenation of ethyl pyruvate in the presence of a platinum catalyst modified with cinchonidine [12], it is not yet clear whether the cinchonidine acts as an adsorbed modifier on the catalyst or as an auxiliary dissolved in the reaction mixture, interacting with the substrate too, and there is no decisive answer on the nature of this interaction.

Accordingly our classification is more or less arbitrary and we did not use strict rules dividing the asymmetric catalytic reactions. The homogeneous chiral metal complex catalysts and reactions were not considered.

DISCUSSION

We take a work done by Corey and co–workers [1] (scheme 1) as first case to discuss.

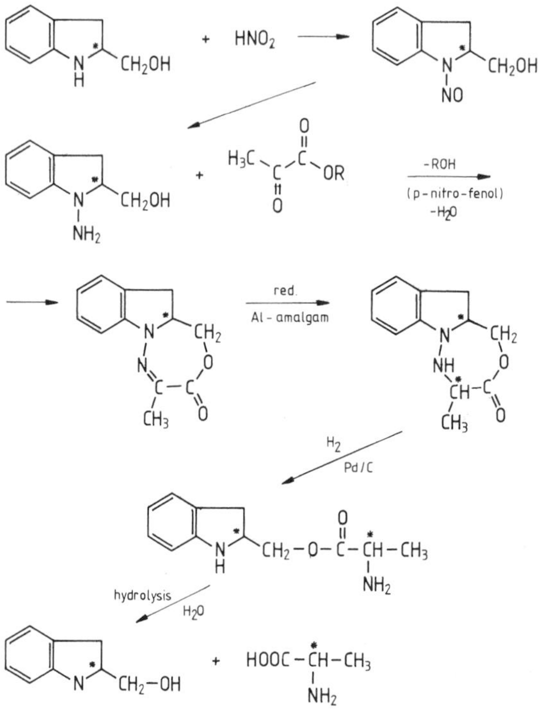

Scheme 1.

2–Indolenyl–methanol was used as chiral source, the 1–amino–derivative was prepared by nitrosation and reduction. This could be reacted with (4–nitro)phenylpyruvate, resulted in a cyclic hydrazone. Its diastereoselective reduction step was carried out with aluminium–amalgam, with 100% stereoselectivity. The next reduction step, the cleavage of the N–N bond was a catalytic hydrogenation with Pd/C. It is probable, that the authors tried the afore mentioned step also with heterogeneous catalytic hydrogenation, but that its stereo-selectivity was much less.

Why is this example an ideal one?
    (i) the enantiomeric excess of the produced alanine is above 95%,
    (ii) the chiral agent, the indolenyl–methanol can be recovered nearly completely,
    (iii) the asymmetric induction takes place in a rigid ring structured molecule.

Why is this example an inconvenient one?
    (i) the reaction step producing the new stereogenic center was not a catalytic hydrogenation,
    (ii) regeneration of the chiral agent requires several steps.

An early and often cited example is the work of Lipkin and Stewart (2) published in 1939 (scheme 2).

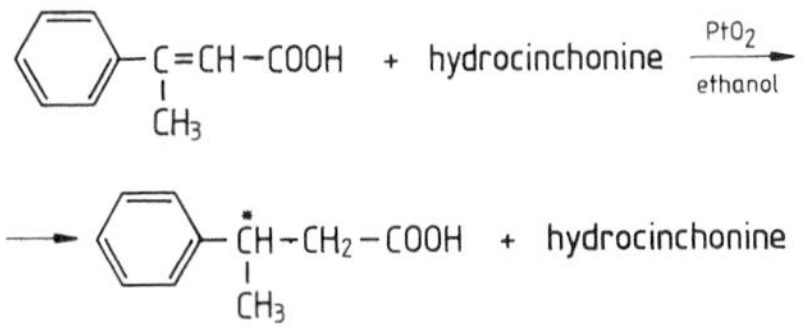

Scheme 2.

They hydrogenated β–methyl–cinnamic acid in the presence of hydrocinchonine in ethanolic solution with platinum oxide. The substrate and the chiral auxiliary form a salt. Because this is a rather weak interaction, the enantiomeric excess was only 9%.

A similar reaction is the hydrogenation of acetophenone oxime with platinum in the presence of menthoxy–acetic acid [3]. The enantiomeric excess was also low (around 9%). See scheme 3.

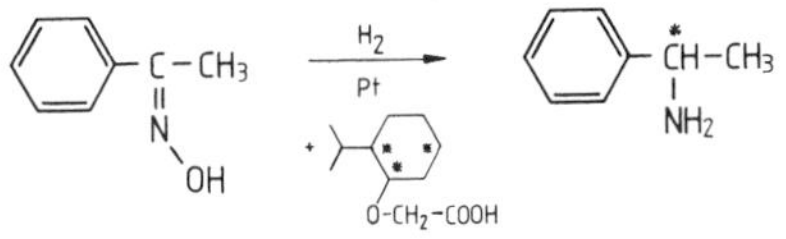

Scheme 3.

Greater enantiomeric excesses were reported (up to 25%) in the nickel catalyzed hydrogenation of diethyl–α–keto–glutaric acid (scheme 4) in the presence of camphor or borneol [4].

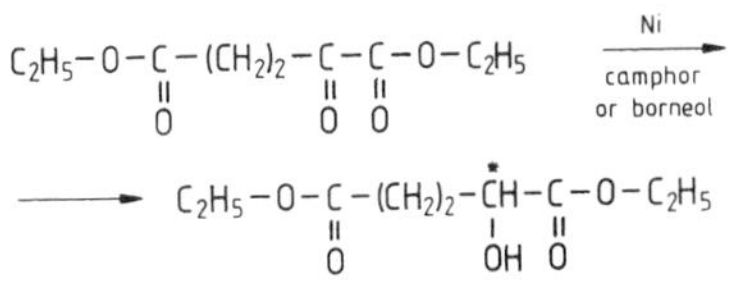

Scheme 4.

The following examples are clearly diastereoselective hydrogenations, where the chiral part of the substrate could be regenerated. Klabunowskii and his coworkers reported on the asymmetric hydrogenation of an azlactone [5], see on the scheme 5.

After hydrolysis phenyl–alanine was obtained in 80 % chemical yield and in 40% ee. The azlactone was first reacted with optically active 1–phenyl–ethyl amine to give α–acetyl–amino–cinnamic acid amide, which was hydrogenated over Pd/C catalyst and hydrolysed into phenyl–alanine and the original amine with good yield.

Horner and coworkers [6] investigated the hydrogenation of methyl–cinnamic acid esters and amides, prepared from menthol and ephedrine as chiral auxiliaries and obtained moderate optical yields. Similar results were achieved in the hydrogenation of keto–acid derivatives (scheme 6).

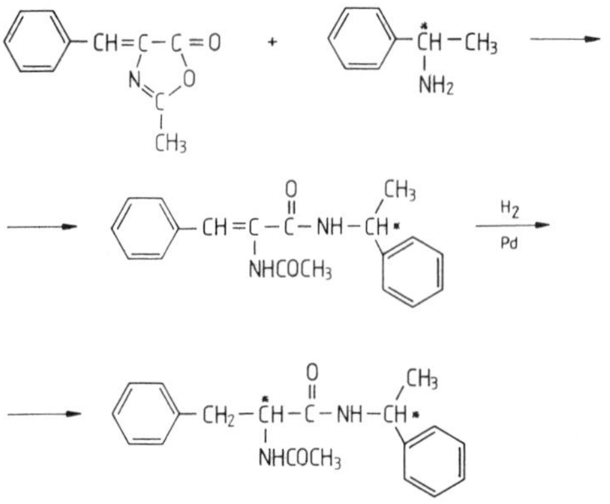

Scheme 5.

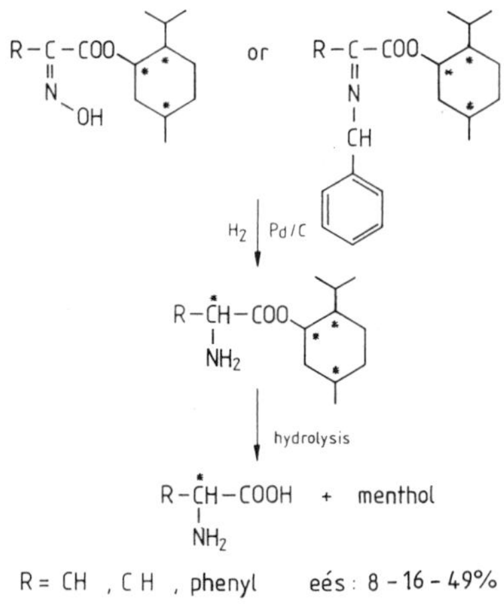

R = CH , $C H$, phenyl   eés : 8 – 16 – 49%

Scheme 6.

More reaction steps were involved in the asymmetric amino–acid synthesis, where the chiral source was also an amino–acid, (S)–proline [7–9].

(S)–proline ester was acylated with pyruvic acid or its derivatives, with dicyclohexyl–carbodiimide as dehydrating agent (scheme 7). The product was left standing with NH_3, to give a hydroxo–diketo–piperazine, which after dehydration and hydrogenation with Pd/C resulted in a diketo–piperazine. Subsequent hydrolysis regenerated the (S)–proline and the product, a new α–amino acid. The optical yield was around 90 %, if R_2 = H, namely the product was alanine, the chemical yield with respect to pyruvic acid reached 60 %. Among the chiral amino acids as reactants, (S)–proline gave the best results.

Similarly, good results were obtained in the reductive amination of pyruvic acid to alanine [10], using (S)–1–ferrocenyl–ethyl–amine as the chiral amine (scheme 8). This chiral molecule could be regenerated after the hydrogenation in two reaction steps. The ferrocenyl–ethyl alanine was cleaved with thio–glyoxilic acid in trifluoro–acetic acid and the resulting thiol reacted with NH_3, to give the optically active ferrocenyl–ethyl amine.

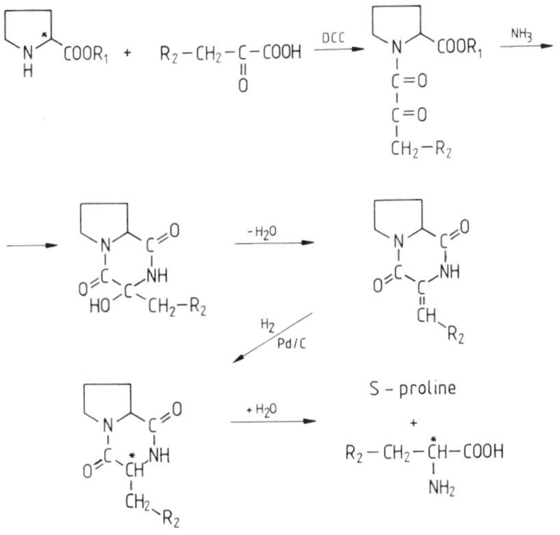

Scheme 7.

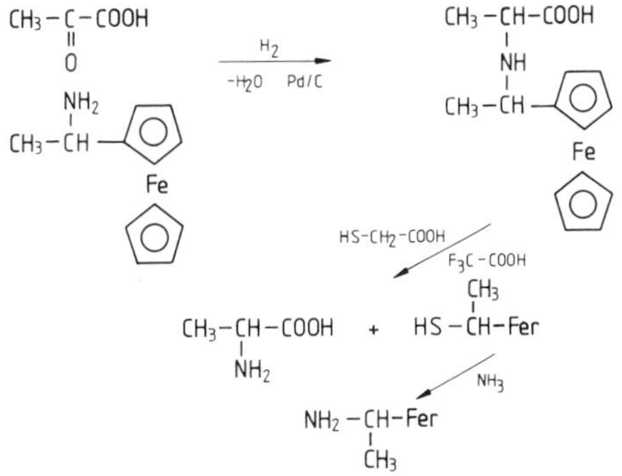

Scheme 8.

Although the vast majority of asymmetric reactions involving heterogeneous catalysts are hydrogenations, asymmetric electrochemical reductions, asymmetric oxidations and other catalytic reactions are known. In reviews written by Blaser and coworkers [11–13] these have been well documented.

We are mentioning two examples: the asymmetric epoxidation of allyl–alcohols [14] with titanium pillared montmorillonite in the presence of diethyl–tartrate (analogy with the homogeneous Sharpless–oxidation [15]) and the asymmetric dihydroxylation of olefins using of a polymer–bound alkaloid and osmium tetroxid catalyst [16].

The chiral modifiers or auxiliaries in these reactions were similar to those applied in hydrogenations: tartaric acid, alkaloids, polypeptides, sugars – all originating from the "chiral pool". The classification of these reactions : whether the catalysts were presumably chiral or the modifier (auxiliary) and the substrate reacted in the solution and a diastereoselective reaction took place, is still a subject of debate.

HYDROGENATIONS IN THE PRESENCE OF (S)–PROLINE

Recently we reported on the hydrogenation of isophorone and acetophenone with a Pd/C catalyst in the presence of (S)–proline [17–22]. The reaction products were optically active dihydroisophorone and 1–phenyl–ethanol. The side products were alkylated prolines (scheme 9).

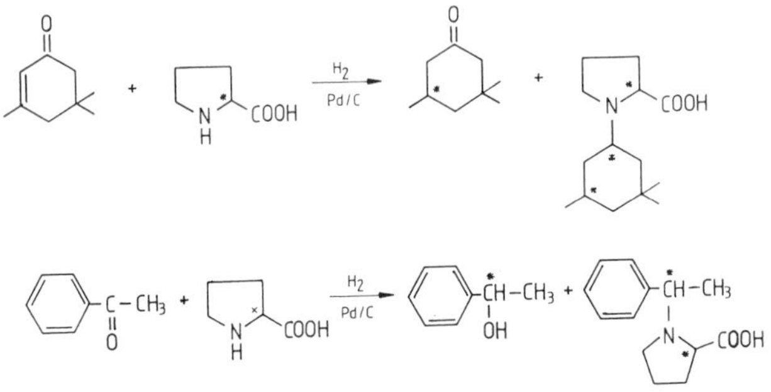

Scheme 9.

During the preliminary experiments we thought, that the chiral "philosophers' stone" had been found, as without prior treatment of the catalyst (merely adding the (S)–proline to the reaction mixture) optically active products were formed. However as more data were collected, we had to conclude, that the optically active products originated from a chemical reaction of the substrates and the chiral auxiliary; resulting in possible intermediates depicted on scheme 10.

For the iminium salt and oxazolidinone formation we have found analogies in the literature [23].

Circular dichroism measurements of the alcoholic solution of (S)–proline and isophorone showed a new peak in the spectra at 286 nm, which increased with time. When the solution of the (S)–proline and isophorone was refluxed for 5–10 minutes before hydrogenation, the optical yields increased and the results were more reproducible.

These observations verified that a chemical reaction took place between the substrate and the chiral auxiliary.

The change of the optical yields with increasing (S)–proline concentration demonstrated the stochiometric relationship between chiral auxiliary and enantiomeric products, as is shown in the figure 1.

As the ratio of (S)–proline to substrate was increased the optical yield increased, but the chemical yield of the saturated product and the amount of regenerated (S)–proline decreased.

Kinetic measurements [22] were carried out in order to make comparisons with other known enantioselective hydrogenations (Raney–nickel/tartaric acid, platinum/cinchonidine). For the latter Garland and Blaser suggested the so–called "ligand–accelerated" model [24], which describes two mechanisms, exhibiting no formal difference : the first is a competitive reaction on modified and unmodified sites, the second is a competitive reaction of the substrate and a possible compound, formed from the substrate and the modifier, on similar catalytic sites. In both cases the reactions of the "modified" substrate or the one on modified sites are faster. In these models the enantioselectivity increases with increasing reaction rates.

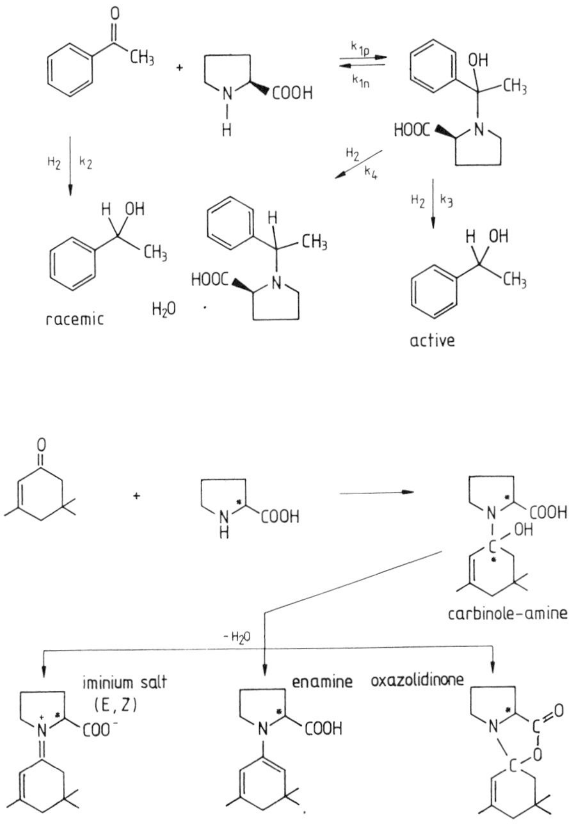

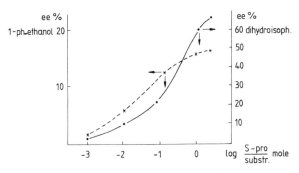

Scheme 10.

Figure 1. Enantiomeric excesses vs. (*S*)–proline : substrate molar ratio.

Klabunovskii [25] suggested a similar kinetic model for the hydrogenation of methyl–aceto–acetate with Raney–nickel modified by tartaric acid. But in this reaction the optical yield was inversely related to the reaction rate, as the latter was lower with modified catalyst than with the unmodified one.

During hydrogenation of reaction mixtures containing isophorone and added (S)–proline, dehydration and reductive reactions are possible (intermediates were shown earlier). Detailed kinetic studies of such a complicated system could not be carried out, because the concentrations of different intermediates could not be determined.

The results of rate measurements are shown in the figure 2.

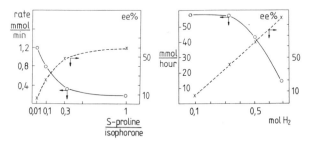

Figure 2. Left : reaction rates (○) and enantiomer excesses (x) as a function of (S)–proline to isophorone molar ratio[a]. Right : rate of hydrogen consumption (○) and ee (x) observed at various conversions expressed in consumed H_2[a,b].
[a] 2 g 10% Pd/C (Selcat) [27], 25 °C, 5 bar, 0.4 mol. isophorone, 0.8 dm^3 methanol.
[b] 0.4 mol proline.

With increasing ratio of chiral auxiliary and conversion, the reaction rate decreases and the enantioselectivity increases. The reaction order with respect to isophorone is not zero. Comparing this with the "ligand–accelerated" model, the enantiomeric excess can not be attributed to competitive reactions taking place on two types of active sites.

A more detailed, but still partial kinetic study was carried out for the hydrogenation of acetophenone–(S)–proline mixtures, we have supposed a reaction scheme (scheme 11) involving addition and hydrogenation reactions.

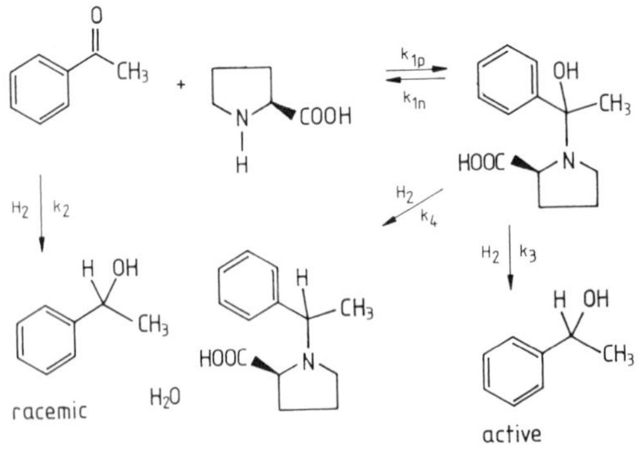

Scheme 11.

Rate measurements have shown that the hydrogenation is of first order with respect to acetophenone. On the figure 3a reaction rates are depicted as a function of conversion with different amounts of (S)–proline:

Figure 3b shows the enantiomeric excesses as a function of conversion; also at different (S)–proline ratios.

In the figure 3c the enantiomeric excesses are shown as a function of reaction rates.

The most important observations are:

(i) the hydrogenation of acetophenone is first order in substrate, in the presence of triethyl–amine or (S)–proline, but is zero order if both triethyl–amine and (S)–proline are present together,

(ii) ee increases slightly with conversion and with increasing (S)–proline ratio but decreases linearly with increasing reaction rates,

(iii) the quantity of (S)–proline recovered from the reaction mixture shows that with increasing initial proline concentration the competing alkylation reaction becomes more significant.

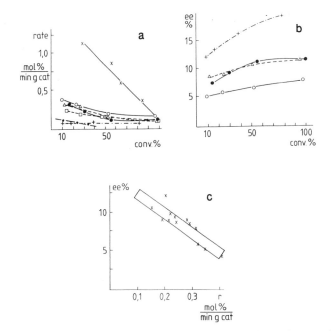

Figure 3a. Reaction rates as a function of conversion in the hydrogenation of acetophenone with increasing (S)–proline/acetophenone ratio : x<Δ<▫<●.

Figure 3b. Enantiomeric excesses as a function of conversion in the hydrogenation of acetophenone with increasing (S)–proline/acetophenone ratio : ○<Δ<● and plus triethylamine (+).

Figure 3c. Enantiomeric excesses vs. reaction rates.

160 mg 10% Pd/C catalyst, Selcat A [27], 25 ℃, 5 bar hydrogen pressure, 0.2 mol. acetophenone, 0.2 dm³ methanol, 4 g triethylamine (+).

The kinetic approach confirmed our assumption that the optically active 1–phenyl-ethanol arose from the hydrogenolytic cleavage of the C–N bond of the adduct of (S)–proline and acetophenone.

The hydrogenations of isophorone and acetophenone in the presence of (S)–proline show similarities: the effect of the Pd/C–(S)–proline system is based on the addition reaction between the reactants and (S)–proline in solution and on the chemoselectivity of Pd. Both hydrogenations should be termed diastereoselective rather than enantioselective, since the asymmetric induction takes place in the hydrogenation of the adduct molecules.

NEW ASYMMETRIC HETEROGENEOUS CATALYTIC HYDROGENATIONS

Because of the apparent limitations of this system, we are trying to find other chiral auxiliaries, substrates and catalysts, which produce enantiomers in excess.

Our approach is based on the following considerations: the chiral compounds should contain secondary or tertiary nitrogen and they should be easily accessible.

As potential chiral auxiliaries, amino–sugars (scheme 12) were synthesized from glucose. First the glucosyl amine was prepared by reacting glucose with the appropriate amine, than it was hydrogenated in methanolic solution with Pd/C catalyst under 40 bar pressure and at 90 °C.

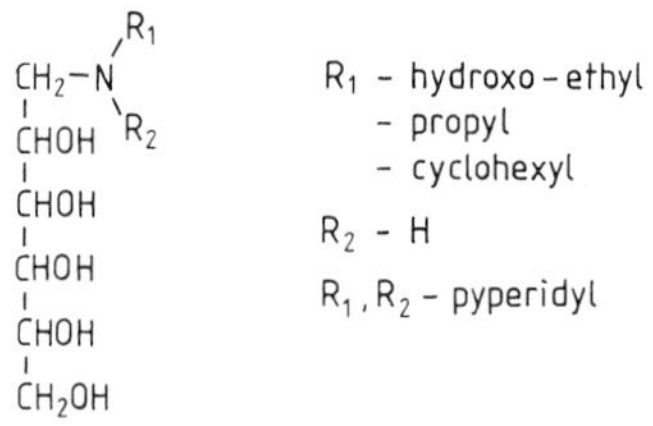

$$
\begin{array}{ll}
\mathrm{CH_2-N}\!\!\overset{R_1}{\underset{R_2}{\diagup}} & R_1 - \text{hydroxo–ethyl} \\
\mathrm{CHOH} & \quad - \text{propyl} \\
\mathrm{CHOH} & \quad - \text{cyclohexyl} \\
\mathrm{CHOH} & R_2 - \text{H} \\
\mathrm{CHOH} & R_1, R_2 - \text{pyperidyl} \\
\mathrm{CH_2OH} &
\end{array}
$$

Scheme 12.

Another chiral auxiliary (scheme 13) was synthesized from +(R)–2–amino–butanol by reductive alkylation. Amino–butanol is the intermediate of an antituberculotic agent, ethambutol.

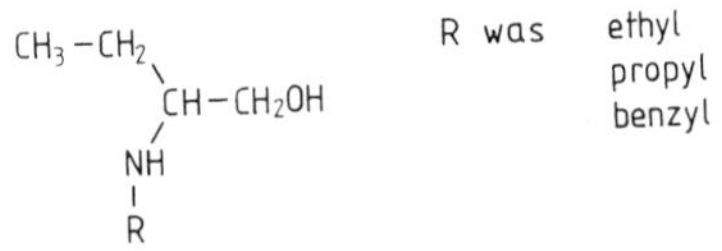

$$
\begin{array}{ll}
\mathrm{CH_3-CH_2}\!\!\diagdown & R \text{ was} \quad \text{ethyl} \\
\qquad\quad \mathrm{CH-CH_2OH} & \qquad\qquad \text{propyl} \\
\qquad\quad \mathrm{NH} & \qquad\qquad \text{benzyl} \\
\qquad\quad \mathrm{R} &
\end{array}
$$

Scheme 13.

These compounds were screened in the hydrogenation of ethyl pyruvate with Pt/C and of isophorone and acetophenone with Pd/C in methanolic solution. The solubility of the sugar amines in methanol is low and the optical yield was also low: 5–6%. The sulphate salts of these compounds have higher solubility, but their chiral effect was very small. The alkylated amino–butanols also had little effect and they underwent competing alkylation by the substrate to a great extent.

The Hungarian pharmaceutical industry produces several kinds of alkaloids, which seemed to be appropriate for our purposes. In the table 1, we summarize the preliminary results of the screening, where for comparison, the results with cinchonidine and (S)–proline under same circumstances are also listed.

Table 1. Enantiomeric excesses (%) in hydrogenations.

Chiral Auxiliaries	Substrates / Catalysts		
	Ethyl pyruvate	Isophorone	Acetophenone
	Pt/C	Pd/C	Pd/C
(S)–Proline	0	2–3	1–2
Cinchonidine	20	3	3
Dihydro–tabersonine	3	6	0
Dihydro–lizergole	4	1,5	4
Dihydro–vinpocetin	30	13	6
		15 [a]	
		20 [b]	
+ vincamine	4	0	0
codein	1	< 1	3,5

[a] Catalyst : Pd/SiO_2.
[b] Catalyst : Pd/TiO_2

Compound name	Formula
dihydro-tabersonine	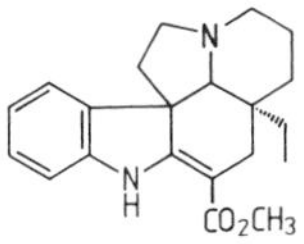
dihydro-lizergole	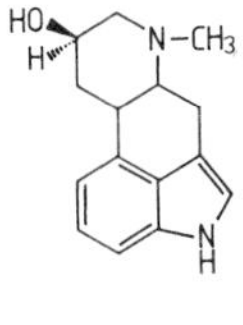
dihydro-vinpocetine (dihydro-apovincaminic acid ethyl ester)	

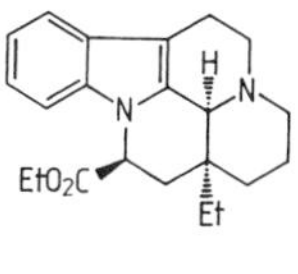

Scheme 14.

The molar ratio of chiral auxiliaries to substrates were between 1:50 and 1:200. The hydrogenations were carried out in methanolic solutions, with 1–5 w% catalysts with respect to the substrate.

The structural formula of the effective compounds can be found on the scheme 14.

It should be mentioned that vinpocetin (apovincaminic acid ethyl ester) is a registered compound of the Richter Gedeon Company, it is a synthetic molecule, which is used to treat oxygen deficiency in the brain [26].

The research work is in progress in order to find the optimum mode, catalyst and parameters for the use of these new chiral auxiliaries. In addition we are screening other chiral molecules, mainly alkaloids from the "chiral pool" to be used as chiral auxiliaries in heterogeneous catalytic hydrogenations.

SUMMARY

A number of asymmetric heterogeneous catalytic reactions are known, where chiral auxiliaries are used in order to produce one enantiomer in excess.

The mode of use of these chiral agents is different:
(i) the most simple way is to simply add them to the reaction mixture,
(ii) prior modification of the catalyst with the chiral agent,
(iii) separate reaction of the substrate with the chiral auxiliary, than to carry out the reaction (for example hydrogenation) forming the new stereogenic center with high diastereoselectivity and finally regenerate the chiral auxiliary.

A common feature of all these reactions is that asymmetric induction takes place in a chiral adduct and this could involve reaction of a chiral auxiliary–substrate adduct with the catalyst or a chiral modifier–catalyst adduct with the (prochiral) substrate or both.

The efficiency of these chiral auxiliaries can be expressed with the enantiomeric excess in the catalytic reaction, the molar ratio of the substrate and the auxiliary and the chemoselectivity and yield of the catalytic reaction.

ACKNOWLEDGEMENTS

This work was supported by the Hungarian OTKA Foundation, the CEC COST Chemistry D2 , Richter Gedeon Comp. and Alkaloida Comp.

REFERENCES

[1]   E.J. Corey, H.S. Sachdev, I.Z. Gougoutas, W. Saenger, J. Am. Chem. Soc., **92**, 8, 2488 (1970).
[2]   D. Lipkin, T.D. Stewart, J. Am. Chem. Soc., **61**, 3295 (1939).
[3]   J. Nakamura, J. Chem. Soc. Jap. **61**, No.5,1051 (1940).
[4]   T. Isoda, A. Ichikawa, T. Shimamoto, J. Sci. Res. Inst. **34**, No.2, 134, 143 (1951).
[5]   A.I. Karpenskaya, L.F. Godunova, E.S. Neukopoeva, E.I. Klabunovskii, Izv. Akad. Nauk. Ser. Chim. **1978**, No.5, 1104.
[6]   L. Horner, H. Ziegler, H.D. Ruprecht, Liebigs Ann. Chem. **1979**, 341.
[7]   B.W. Bycroft, G.L. Lee, J. Chem. Soc. Chem. Commun. **1975**, 988.
[8]   N. Izumiya, S. Lee, K. Kanmera, H. Aoyagi, J. Am. Chem. Soc, **99**, 8346 (1977).
[9]   T. Kanmera, S. Lee, H. Aoyagi, N. Izumiya, Int. J. Peptide Protein Res. **16**, 280 (1980).
[10] A. Ratajczak, A. Czech, Bull. Acad. Pol. Sci Chim., **27**, 661 (1979).
[11] H.U. Blaser, M. Müller, Studies in Surf. Sci. and Cat., Heterogeneous Catalysis and Fine Chemicals II (Guisnet et al. Eds.) Elsevier (1991), p 73.
[12] H.U. Blaser, Tetrahedron Asymmetry **2**, No.9, 843 (1991).
[13] H.U. Blaser, Chem. Rev. **1992**, 92, 935.
[14] B.M. Choudary, V.L.K. Valli, A. Durga Prasad, J. Chem. Soc. Chem. Commun. **1990**, 1186.
[15] K.B. Sharpless, Chemtech. **1985**, 692.

[16] B. Moon Kim, K.B. Sharples, Tetrahedron Lett., **31**, No. 21, 3003 (1990).

[17] A. Tungler, M. Kajtár, T. Máthé, G. Tóth, E. Fogassy, J. Petró, Catalysis Today, **5**, (1989) 159.

[18] A. Tungler, T. Máthé, J. Petró, T. Tarnai, J. Mol. Catal., **61**, (1990) 259.

[19] A. Tungler, T. Tarnai, T. Máthé, J. Petró, J. Mol. Catal., **67**, (1991) 277.

[20] A. Tungler, T. Tarnai, T. Máthé, J. Petró, J. Mol. Catal., **70**, (1991) L5.

[21] G. Tóth, A. Kovács, T. Tarnai, A. Tungler, Tetrahedron Asymmetry, Vol.**4,** No. 3 (1993) 331.

[22] A. Tungler, T. Tarnai, A. Deák, S. Kemény, A. Gyôry, T. Máthé, J. Petró, Studies in Surf. Sci. and Cat. Heterogeneous Catalysis and Fine Chemicals III (Guisnet et al. Eds.) Elsevier (1993), p 99.

[23] M. Joucla, J. Mortier, Bull. Soc. Chim. France, **1988**, No.3, 579.

[24] M. Garland, H.U. Blaser, J. Am. Chem. Soc., **112**, 748 (1990).

[25] E.I. Klabunovskii, A.A. Vedenyapin, Asym. Kataliz, Gidrog. na Metal. Nauka, Moscow, (1980).

[26] Richter Gedeon Comp. US P. 4 035 370

[27] T. Máthé, A. Tungler, J. Petró, US P 4 361 500.

ENZYMATIC RESOLUTION OF BIOLOGICALLY ACTIVE PRECURSORS VERSUS ASYMMETRIC INDUCTION IN HETEROGENEOUS CATALYSIS

H. Hönig, R. Rogi-Kohlenprath and H. Weber

Institute of Organic Chemistry
Graz University of Technology
Stremayrgasse 16
A–8010 Graz, Austria

INTRODUCTION

Since the first isolation of the potent immunosuppressor FK 506 [1,2] (figure 1), many different approaches appeared in the literature for the synthesis of its cyclohexyl moiety. Some strategies, especially for the enantiomerically pure fragment are brilliant pieces of organic synthesis, although somewhat elaborate [3]. The average number of steps equals ten, some approaches using enantiopure educts. The classical way to such cyclohexyl fragments are Diels–Alder–Reactions [4]. Some of the more than twenty papers on the syntheses of structures like **3** use this methodology. To our knowledge, up to now, there is no report on the obviously most simple approach: hydrogenation of vanillic acid or derivatives thereof in order to obtain racemic products like **1**. Inversion of configuration at C–4 by standard methods should provide the racemic compound **2**, which easily can be separated from the undesired enantiomer by enzymatic resolution yielding enantiopure **3** [5] (scheme 1).

As another example, structures like **4**, intermediates in the synthesis of some trichotecenes, were obtained by yeast reduction with low e.e. [6] (scheme 2). Again, the simple hydrogenation of an accordingly substituted aromatic compound, which in this case is commercially available, is also not yet described in the literature.

The fact, that the most obvious ways to such compounds have not yet been tried, might be due to two results known from literature [7] :

1) Methoxy– and hydroxy substituents tend to be cleaved by hydrogenolysis with common catalysts like platinum or nickel.
2) Hydrogenations of aromatics quite often are not that highly stereospecific *all–cis* as normally expected.

As the target structures also exhibit chirality, from a theoretical standpoint, asymmetric hydrogenation with the aid of a chiral auxiliar, although not yet desribed for hydrogenations of aromatics, could further facilitate these syntheses. Here we describe our attempts towards this goal.

Chiral Reactions in Heterogeneous Catalysis
Edited by G. Jannes and V. Dubois, Plenum Press, New York, 1995

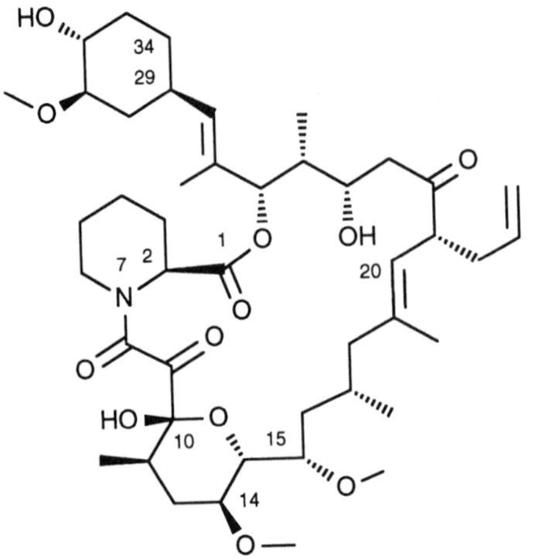

Figure 1. Structure of FK 506.

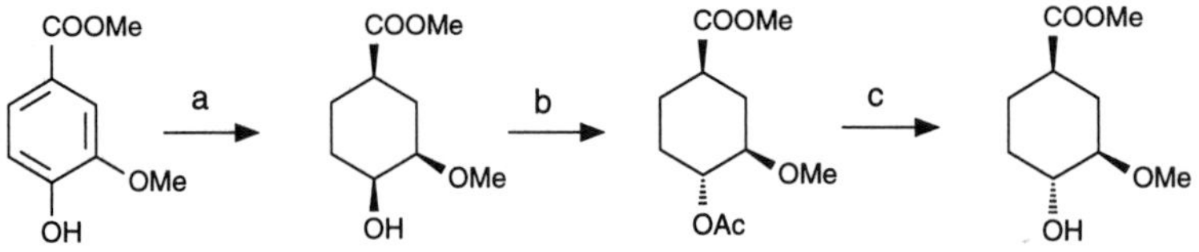

rac-1   rac-2   (R,R,R)-3

a: H_2, 5% Rh/Al_2O_3, MeOH/HOAc, 130 bar, 100°C, 20 h, 65 % (isolated)

b: 1. Tf_2O, pyridine, CH_2, =°C, 35 min;

   2. Bu_4NOAc, CH_3CN, 0°C, 3 h; 40 % (two steps, isolated)

c: enzymatic resolution

Scheme 1. Improved synthesis of **3**.

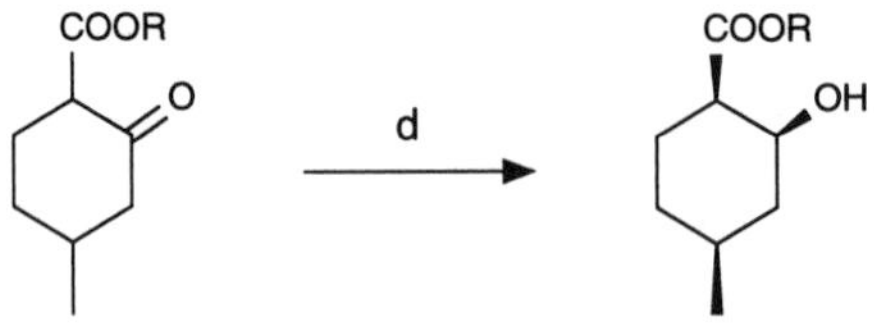

4

d: bakers yeast, sugar, water, 40 %, 45 % e.e.

Scheme 2. Synthesis of **4** as in [6].

RESULTS AND DISCUSSION

As shown in schemes 3 and 4, high pressure hydrogenation over rhodium on alumina of the simple aromatic precursors for **3** and **4** yield different amounts of the desired *all–cis–* products [10] and also differ in their respective stereoselectivity with regard to *cis– trans –* isomers. The assignments of the stereochemistry and identity of the various products obtained was done with g.l.c. and ^{13}C–nmr. In the case of **3**, neither hydrogenolysis of hydroxy– or methoxy–groups nor low stereoselectivity hampers the very straightforward synthetic strategy. From the results obtained, it is obvious, that both stereoselectivity as well as degree of hydrogenolysis very much depends on the nature of the substituents on the aromatic ring. The synthesis of **3** (scheme 1) thus is a fortunate example of this strategy. But in the case of **4**, although not as selective, separation by column chromatography also yields the pure isomers needed.

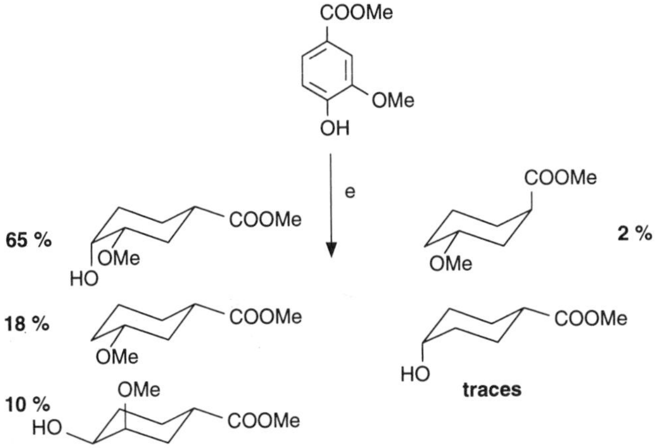

e: H_2, 5 % Rh/Al$_2$O$_3$, MeOH/HOAc, 130 bar, 100°C, 15 h

Scheme 3. Results of the hydrogenations leading to **3**.

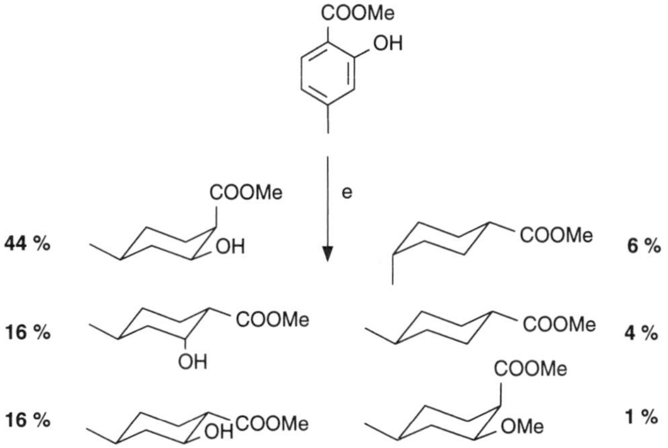

Scheme 4. Results of the hydrogenations leading to **4**.

Rhodium on alumina can be substituted by rhodium on charcoal, while the attempted low pressure hydrogenation with rhodium on strontium titanate described in the literature [8], at least in our hands, was not successful.

The respective products of course all are racemic, but can be converted to enantiopure building blocks by the known method of enzyme catalyzed resolution of the respective acylates of the secondary alcohols [5]. Thus, via the inversion of configuration at C–4 by phase transfer catalyzed exchange of the respective triflate of rac–1, the most simple synthesis of 3 could be achieved.

The butyrate of 1 itself can be hydrolyzed by lipase AK (AMANO [9]). The alcohol shows $[\alpha]_D^{22} = + 16.7$ (c = 1.3, CH_2Cl_2, 97% e.e., 34% conversion in 4.5 d). Hydrolysis with lipase from *Candida rugosa* is somewhat quicker (19 h, 41% conversion), but with lower e.e. (92%). The remaining butyrate shows (after 64% conversion) $[\alpha]_D^{22} = +20.3$ (c = 1.8, CH_2Cl_2, e.e. = 98%).

As mentioned above, the success in the stereoselectivity of the hydrogenations prompted us to investigate the possibility of enantioselectivity in catalytic hydrogenations of aromatic compounds. Although the topological differentiation of the two faces of an unsymmetrically substituted aromatic ring in principle should be possible, the preliminary results obtained by adding different chiral auxiliaries like optically active amines, amino acid derivaties and cyclodextrines to the reaction did not result in any detectable chirality of the hydrogenated products. We will pursue this idea still further by modifying the catalyst with some sterically more demanding chiral "hosts" for the aromatic substrate. Still, the high pressure and temperatures applied could pose some problems in this respect, and therefore we will also pursue the adaption of the low pressure hydrogenation with different catalysts to our substrates.

REFERENCES AND NOTES

[1] Tanaka, H., Kuroda, A., Marusawa, H., Hatanaka, H., Kino, T., Goto, T., Hashimoto, M., Taga, T., *J. Am. Chem. Soc.* 1987, *109*, 5031.
[2] Fagiuloli, S., Gasbarrini, A., Azzarone, A., Francavilla, A., Van Thiel, D.H., Ital. J. Gastroenterol. 1992, 24, 355.
[3] See inter alia :
     (a) Bartlett, P. A., McQuaid, L.A., J. Am. Chem. Soc. 1984, 106, 7854.
     (b) Jones, T.K., Reamer, R.A., Desmond, R., Mills, S.G., J. Am. Chem. Soc. 1990, 112, 2998.
     (c) Linde II, R.G., Egbertson, M., Coleman, R.S., Jones, A.B., Danishefsky, S.J., J. Org. Chem. 1990, 55, 2771.
     (d) Schreiber, S.L., Smith, D.P., J. Org. Chem. 1989, 54, 9.
     (e) Pearson, A.J., Roden, B.A., J. Chem. Soc., Perkin. Trans I 1990, 723.
     (f) Kuhn, T., Tamm, C., Riesen, A., Zehnder, M., Tetrahedron Lett. 1989, 30, 693.
     (g) Chini, M., Crotti, P., Macchia, F., Pineschi, M., Flippin, L.A., Tetrahedron 1992, 48, 539.
     (h) Thom, C., Kocienski, P., Jarowicki, K., Synthesis 1993, 475.
[4] Ireland, R.E., Highsmith, T.K., Gegnas, L.D., Gleason, J.L., J. Org. Chem. 1992, 57, 5071.
[5] Gu, R.L., Sih, C.J., Tetrahedron Lett. 1990, 31, 3287.
[6] Gilbert, J.C., Selliah, R.D., Tetrahedron Lett. 1992, 33, 6259.
[7] Rylander, P. N., *Catalytic Hydrogenation over Platinum Metal,* Acad. Press, New York, 1967, 338–45
[8] Timmer, K., Thewissen, D.H.M.W., Meinema, H.A., Bulten, E.J., Recl. Trav. Chim. Pays–Bas 1990, 109, 87.
[9] Donations by AMANO are gratefully acknowledged.
[10] Since we could not find any ^{13}C NMR data of compounds similar to 1 to 4, the respective values are given below (CDCl$_3$, in ppm, 75 MHz) :

   1: 175.6 (COOR); 80.1 (C3); 66.0 (C4); 56.1 (OMe); 51.6 (COOMe); 40.9 (C1); 29.3 (C5*); 27.7 (C2*); 22.3 (C6)

3: 175.1 (COOR); 84.1 (C3); 73.3 (C4); 56.8 (OMe); 52.0 (COOMe); 41.7 C1); 31.1 (C2*); 31.0 (C5*); 26.9 (C6)

4: 174.9 (COOR); 70,6 (C2); 51.4 (C1); 45.2 (COOMe); 40.2 (C3); 31.4 (C5); 30.4 (C4); 26.6 (C6); 21.9 (Me)

ENANTIO- AND DIASTEREOSELECTIVE REDUCTION OF DISUBSTITUTED AROMATICS

K. Nasar, M. Besson*, P. Gallezot, F. Fache and M. Lemaire

Institut de Recherches sur la Catalyse–CNRS
2, Avenue Albert Einstein
69626 Villeurbanne Cedex (France)
and Laboratoire de Catalyse et Synthèse Organique
Université Claude Bernard–Lyon 1– ESCIL
69622 Villeurbanne Cedex (France)

* Author to whom correspondance should be addressed at the Institut de Recherches sur la Catalyse

INTRODUCTION

Biological activity is closely tied to chirality and control of the enantioselectivity in the synthesis of fine chemicals is of increasing interest [1–3]. Although a large number of enantioselective reductions of prochiral compounds are reported in the literature, no catalytic enantioselective reductions of disubstituted aromatics, which could lead to valuable alicyclic compounds have been reported yet. To reach this goal, two approaches were tried in the present work, namely enantioselective and diastereoselective heterogeneous catalysis.

In enantioselective catalysis, the surface of the metal catalyst is modified with an optically pure compound. Few heterogeneous catalysts are known to give enantioselective hydrogenation. Nickel catalysts modified by (R)–(+) tartaric acid have been used early for asymmetric hydrogenation of ß–ketoesters and ß–diketones [4]. More recently, the enantioselective reduction of a–ketoesters was achieved with platinum catalysts modified by cinchona alkaloids [5]. We have recently succeeded in reducing dibenzo–18–crown–6 ether into the cis–syn–cis isomer of dicyclohexyl–18–crown–6 ether with a 95% selectivity using rhodium under phase transfer conditions [6]. The catalyst consists in small colloidal rhodium particles stabilized with amines or ammonium salts. This catalytic system was then used in the hydrogenation of 2–methylanisole [7]. In the present work, we attempted to extend its use to the asymmetric reduction of substituted aromatics using chiral amines as inductor.

In diastereoselective catalysis, chiral auxiliaries are anchored to the substrate to be hydrogenated. These chiral sources available from the chiral pool [8] are mainly derived from natural amino–acids or terpenes. These chiral auxiliaries controlling the stereochemistry should be easily removed after reduction without damaging the hydrogenated substrate. In this work we have used (–) menthoxyacetic acid and different (S)–proline derivatives as chiral auxiliaries for the reduction of ortho–cresol and ortho–toluic acid respectively. Proline derivatives have been used previously as auxiliaries for diastereoselective reductions. Thus, the (S)–proline tertiobutyl ester was used in the asymmetric hydrogenation of dihydropeptides and the optical purity of the resulting alanine was 40–50%. The use of (S)–proline tertiobutylamide with dehydrotripeptides yielded *ee* in the range 43 to 93% [9]. (S)–proline esters or amide were also used as the chiral source in the hydrogenation of pyruvamides. (S)–proline isopropylamide was a more effective chiral moiety than the other (S)–proline derivatives [10–11]. The reductions of ortho–cresol and ortho–toluic acid anchored to the chiral auxiliaries were carried out either on the colloidal catalytic system or on a Rh/C catalyst, with or without modifiers, and in different solvents.

EXPERIMENTAL

Materials and Synthesis of Substrates

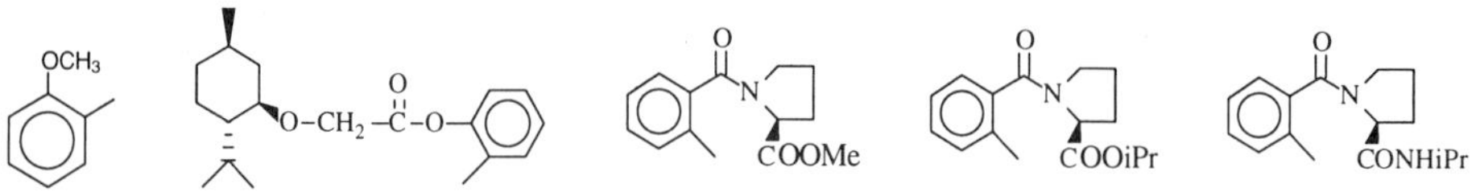

Scheme 1.

2–methylanisole (**1**) was purchased from Aldrich and distilled under vacuum before use. Ortho–cresylmenthoxyacetate (**2**) was prepared by condensation of menthoxyacetic acid chloride with o–cresol in presence of triethylamine (yield after purification : 84%). To prepare compounds (**3–4**), (S)–proline methyl and isopropyl ester chlorhydrates, obtained by esterification of (S)–proline with $SOCl_2$ in methanol or isopropanol, were condensed with o–toluoylchloride in presence of triethylamine in $CHCl_3$ (yield after purification : 73%). To prepare substrate (**5**), (S)–proline and o–toluoylchloride were condensed. The carboxylic function is then derived into the isopropyl amide by peptidic coupling with isopropylamine in presence of dicyclohexylcarbodiimide (DCC) and of N–hydroxy-succinimide (yield after purification : 58%).

The synthesis of (R)–(–)–N,N–dioctylcyclohexyl–1–ethylamine (DOCEA) was performed by reacting (R)–(–)1–cyclohexylethylamine and n–octanal in ethanol in presence of Pd/C catalyst under 5 MPa hydrogen pressure. The other amines were commercial and were distilled before use.

Preparation of the Catalysts and Hydrogenation Methods

The reductions of 2.4 mmol of substrates **1–5** were performed at room temperature, under 5 MPa hydrogen pressure in a 30 ml magnetically stirred autoclave, using either a commercial 5% Rh/C catalyst (Aldrich) in 15 ml solvent or colloidal rhodium prepared in-situ by the following procedure. A 5 ml aqueous solution of 0.24 mmol of $RhCl_3.H_2O$ was mixed with 10 ml CH_2Cl_2 containing the amine and the substrate. The reactor was flushed with argon. After introduction of 5 MPa hydrogen and stirring, the rhodium was reduced and the reaction started. Sampling of the reaction medium could be performed during the run.

Analytical Methods

The chemical yields of all reductions were measured, using a J&W DB–1701 column. The enantiomeric excesses obtained after hydrogenation of 2–methylanisole (1) were measured by polarimetry (Perkin–Elmer 241) and by chiral gas chromatography on a SGE CYDEX–ß column.

The two cis diastereoisomers obtained by hydrogenation of the o–cresyl(–) menthoxyacetic ester (2) could not be separated on the DB–1701 column (this was demonstrated by direct synthesis of the hydrogenated products from menthoxyacetic acid chlorid and cis–2–methylcyclohexanol). Therefore, the reaction products were hydrolysed in MeOH/HCl under reflux into 2–methylcyclohexanols then transformed into 2 methyl–1–methoxy–cyclohexanes which were analysed by chiral chromatography on the CYDEX–ß column.

The hydrogenated products of substrates (3–5) were separated on the DB–1701 column. Positive diastereoisomeric excesses (de >0) mean an excess of the second peak of two diastereoisomers. The enantiomeric excesses (ee) were determined on a Macherey–Nagel Düren LIPODEX column after hydrolysis into 2–methyl–1–cyclohexanecarboxylic acids and esterification.

RESULTS AND DISCUSSION

Enantioselective Reduction

Various chiral amines, playing the role of both transfer agent and chiral inductor for rhodium, were used in the reduction of 2–methylanisole in the biphasic water–dichloromethane system. No chiral induction was observed with (–)–cinchonidine, (S)–(–)–nicotine, (–)–sparteine or (R)–(–)–1–cyclohexylethylamine for instance. With (R)–(–)–N,N–dioctylcyclohexyl–1–ethylamine (DOCEA) an enantiomeric excess of ca. 3–6% was observed with an amine/Rh molar ratio of 3.5. The same amine or N–methyl–(S)–prolinol associated to a Rh/C catalyst in dichloromethane leads to the same enantiomeric excesses.

Diastereoselective Reduction

The results of the catalytic hydrogenations are summarized in table 1.

Reduction of o–cresol with menthoxyacetic acid auxiliary. Hydrogenation of the o–cresyl(–)menthoxyacetic ester (2) gives only two cis–products with the auxiliary still attached. After hydrolysis, the two corresponding 2–methylcyclohexanols are obtained in ca. 10% enantiomeric excess, whether in presence of a modifier or not. However, due to the acidic nature of the hydrolysis medium, an epimerization of the enantiomers may occur so that the initial diastereoisomeric excesses might be higher. Work is in progress to improve the determination of the de and ee of the hydrogenated products.

Reduction of o–toluic acid with proline derivatives auxiliaries. Upon reduction of substrates **3–5** , the two cis–hydrogenated isomers are by far the major products (>95%). Diastereoisomeric excesses (*de*) were calculated taking into account these two cis–products. Using either colloidal rhodium or Rh/C without addition of modifiers, the hydrogenated products are obtained with a negative *de*. However, in presence of modifiers an inversion of the diastereoselectivity is observed.

Table 1. Diastereoselective catalytic hydrogenation of substrates **2–5**.

Substrate	Catalyst precursor	Modifier	Mod/Rh (molar ratio)	Solvent	Conversion (%)	de[a] (%)
2	Rh/C	-	0	CH_2Cl_2	100	13[c]
	colloidal Rh	TOA[d]	3	CH_2Cl_2/H_2O	100	10[c]
3	colloidal Rh	-	0	CH_2Cl_2/H_2O	93	-12
		TOA[d]	3.5	CH_2Cl_2/H_2O	100	10
		TPA[e]	3.4	CH_2Cl_2/H_2O	100	14
		(-)DOCEA[f]	3.4	CH_2Cl_2/H_2O	90	35
	Rh/C	-	0	iPrOH	100	-13
		(-)DOCEA[f]	3.5	iPrOH	77	33[b]
		(+)DOCEA[f]	3.5	iPrOH	100	28
		EDCA[g]	3.5	iPrOH	100	37
		(-)sparteine	3.4	iPrOH	100	31
		-	0	MeOH	100	-14
		(-)DOCEA[f]	3.7	MeOH	43	42
4	Rh/C	-	0	iPrOH	100	-22
		(-)DOCEA[f]	3.5	iPrOH	100	25
5	Rh/C	-	0	iPrOH	100	-20
		(-)DOCEA[f]	3.5	iPrOH	100	6
	Rh/C	-	0	MeOH	97	-13
		(-)DOCEA[f]	3.5	MeOH	47	16

a *de* >0 means an excess of the second peak of the hydrogenated product

b *ee* of the product freed from the auxiliary was determined by chromatography and was found to be 30%

c *ee* after subsequent hydrolysis determined by chromatography(*ee* =13%) or by polarimetry (*ee* = 10%)

d TOA = trioctylamine

e TPA = tripentylamine

f DOCEA = dioctylcyclohexyl-1-ethylamine

g EDCA = N-ethyldicyclohexylamine

Upon reduction of 1–(2–methylbenzoyl)–2–(methyl ester)–(S)–proline (**3**), the colloidal rhodium stabilized with trioctylamine (TOA) or tripentylamine (TPA) gives 10% and 14% *de* respectively. Higher *de* are obtained using (–) DOCEA (*de* =35%). With the same amine, the commercial Rh/C catalyst gives comparable induction either in isopropanol (de=33%) or in methanol (de=42%). The configuration of the amine is not determining since (+) DOCEA and (–) DOCEA give 28% and 33% de respectively in isopropanol. In agreement, the achiral N–ethyldicyclohexylamine (EDCA) leads to a 37% de. Sparteine is also an efficient modifier (de=31%).

Reduction of 1–(2–methylbenzoyl)–2–(isopropyl ester)–(S)–proline (4) gives de =25%. This shows that substitution of the methyl group in the (S)–proline ester by the more bulky isopropyl group results in a slight decrease of the asymmetric induction (de=25% instead of 33%).

Finally, using proline isopropyl amide (5) as auxiliary, 6% and 13% de are obtained in isopropanol and methanol respectively. This group is as bulky as the corresponding ester, but has a more rigid structure. Nevertheless, it was a much less effective chiral auxiliary .

CONCLUSIONS

Enantioselective excesses after reduction of 2–methylanisole with colloidal rhodium stabilized and modified by a chiral amine are weak. In contrast, the chiral induction is significant during the hydrogenation of substituted aromatics containing various chiral auxiliaries. The nature of the anchored group is important and successfull asymmetric induction was obtained with the methyl (S)–proline ester in presence of DOCEA or EDCA (de ca. 40%). The efficiency of these chiral auxiliaries can be explained by a chelation mechanism, in agreement with previous studies using (S)–proline moieties [1,10–11]. It is assumed that the polar N and O atoms of the substrate adsorb on the catalyst surface to form a tightly bound substrate–catalyst surface complex, then the aromatic moiety adsorbs on the catalyst from the less bulky side of the molecule. However in the present study an improvement in diastereoselectivity and a configurational inversion is observed in presence of an amine, the role of these modifiers should be clarified in future work.

REFERENCES

[1]   Asymmetric Synthesis, Vol 5 : Chiral Catalysis, (J.D. Morrison Ed.), Academic Press (1985).

[2]   H.U. Blaser, Tetrahedron : Asymmetry, 2(9), 843 (1991).

[3]   R.A. Sheldon, Chirotechnology, Industrial Synthesis of Optically Active Compounds, Marcel Dekker (1993).

[4]   Y. Izumi, Advances Catal., 32, 215 (1983).

[5]   H.U. Blaser, H.P. Jalett, D.M. Monti, J.F. Reber and J.T. Wehrli in Heterogeneous Catalysis and Fine Chemicals, (M. Guisnet Ed.), Elsevier, 153 (1988).

[6]   P. Drognat–Landre, D. Richard, P. Gallezot and M. Lemaire, J. Mol. Cat. , 78, 257 (1993).

[7]   K. Nasar, F. Fache, M. Lemaire, M. Besson and P. Gallezot, J. Mol. Cat. , 87, 107 (1994).

[8]   J. Crosby, Tetrahedron, 47(27), 4789 (1991).

[9]   K. Harada, Asymmetric Synthesis, Vol 5, 345 (1985).

[10] T. Munegumi, M. Fujita, T. Maruyama, S. Shiono, M. Takasaki and K. Harada, Bull. Chem. Soc. Jpn. , 60, 249 (1987).

[11] T. Munegumi, T. Maruyama, M. Takasaki and K. Harada, Bull. Chem. Soc. Jpn. ,63, 1832 (1990).

HETEROGENEOUS ASYMMETRIC CATALYSIS WITH C$_2$ SYMMETRIC AMINE–MODIFIED RHODIUM/SILICA

P. Gamez[1], F. Fache[1], M. Lemaire[1] and P. Mangeney[2].

[1] Institut de Recherches sur la Catalyse
Laboratoire de Catalyse et Synthèse Organique, UP CNRS 5401,
Université Claude Bernard Lyon I, ESCIL, Bât. 308,
43 bd du 11 novembre 1918,
F– 69622 Villeurbanne CEDEX, France.

[2]Laboratoire de Chimie des Organo–éléments,
Université P. et M. Curie, tour 45,
4 place Jussieu,
F–75252 Paris CEDEX 05, France.

INTRODUCTION

Supported homogeneous catalysis is generally obtained by synthesis of material in which ligand is covalently bound to the support [1]. Such synthesis are generally neither straightforward nor easy to reproduce.

In this paper, we report our first results on the heterogeneous catalytic reduction of ketones into their corresponding alcohols. The catalytic site is formed by Rhodium (I) and C$_2$ symmetric chiral diamines as ligands, bound to a SiO$_2$ support by non covalent interactions, in a continuous flow reactor (figure 1).

EXPERIMENTAL

The chemical yields were measured on a capillary chiral column Macherey–Nagel–Düren, Lipodex E (25 m, $\varnothing$ = 0.25 mm); enantiomeric excesses were measured both by GC and polarimetry.

Batch conditions and related results have been already published [2].

Continuous flow reactor conditions : [Rh(C$_6$H$_{10}$)Cl]$_2$ (22 mg, 0.1 mmol Rh(I)) and the diamines (0.2 mmol) are diluted in 10 ml of solvent. After few minutes, the support (2 g) is added and the mixture is stirred for 1 h under nitrogen. The heterogeneous solution is then filtered off to dryness in the reactor and a solution of the substrate and KOH in pellets

(KOH/Rh = 6) is allowed to pass through the reactor at a flow rate of 0.5 ml.h^{-1}. The filtrate is analysed by GC.

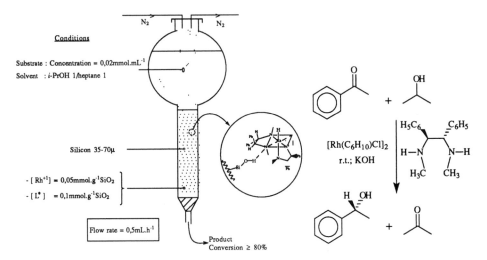

Figure 1. Continuous flow reactor for the enantioselective reduction of prochiral ketones.

RESULTS AND DISCUSSION

We have tested our catalytic system –rhodium and diamines– in homogeneous phase (batch), for comparison (see table 1). Good to excellent e.e. have been obtained (up to 99% e.e. at 100% conversion for the methylbenzoylformate, entry 6). Three different supports have been tested (silica gel 60, 15–40 μm Merck; silica 35–70 μm Amicon; neutral alumina Brockman 1 150 μm Aldrich), but only the silica of 35–70 μm granulometry gave satisfactory results.

Transposition of the batch experiments to the continuous flow reactor leads to the same results. But, an overall turnover number of 300 is observed (entry 7) which has to be compared to the 20 of the homogeneous process (entry 6). Furthermore, at higher dilution, noticeable increase of the e.e. is obtained (entries 2 and 3). A detailed mechanistic study is necessary to understand this effect of the dilution on the e.e.. Nevertheless, the possibility of working at high dilution is one of the advantages of continuous reactor (heterogeneous catalysis) when compared to batch.

Table 1. Comparative results in batch or in continuous flow reactor.

			Batch	Reactor	Continuous	Flow
Entry	R	Solvent	Conversion %	e.e. % (Configuration)	Conversion %	e.e.% (Configuration)
1	CF₃	*i*–PrOH	100	33(R)	/	/
2	CF₃	*i*–PrOH 1/ heptane 1	/	/	100	21(R)
3	CF₃	*i*–PrOH 1/ heptane 1	/	/	100	27(R)[a]
4	CF₃	*i*–PrOH 1/ heptane 1	100	16(R)		
5	CH₃	*i*–PrOH	100	67(R)	80	61(R)
6	COOMe	*i*–PrOH	100	_99(R)	/	/
7	COOMe	*i*–PrOH 1/ heptane 1	/	/	90	_99(R)[b]

a : 0.0067M; **b** : [substrate]/[Rh] = 300

REFERENCES

[1]. A. Corma, M. Iglesias, C. del Pino, F. Sanchez; *J. Chem. Soc., Chem. Commun.*, 1253 (1991).
[2]. P. Gamez, F. Fache, P. Mangeney, M. Lemaire, submitted to *Tetrahedron Lett.*.

AN INVESTIGATION OF THE MODE OF HYDROGEN ADDITION IN STEREOSELECTIVE HETEROGENEOUS CATALYTIC HYDROGENATIONS

A.Tungler[(1)], T.Tarnai[(2)], T.Máthé[(2)], G.Tóth[(3)], J.Petró[(1)] and R.A.Sheldon[(4)]

[(1)]Department of Organic Chemical Technology, Technical Univ. of Budapest
Müegyetem P.O. Box 91
H–1521 Budapest, Hungary
[(2)]Organic Chemical Research Group of the Hungarian Academy of Sciences
Technical University
H–1521 Budapest, Hungary
[(3)]Technical Analytical Research Group of the Hungarian Academy of Sciences
Institute for General and Analytical Chemistry, Technical University
St. Gellért tér 4
H–1111 Budapest, Hungary
[(4)]Laboratory for Organic Chemistry and Catalysis, Delft University of Technology
P.O. Box 5045
NL–2600 GA Delft, The Netherlands

INTRODUCTION

The exact mode of addition of hydrogen [1] to C=C and C=N double bonds in liquid phase heterogeneous catalytic hydrogenations is still an unresolved issue. The various possibilities are:

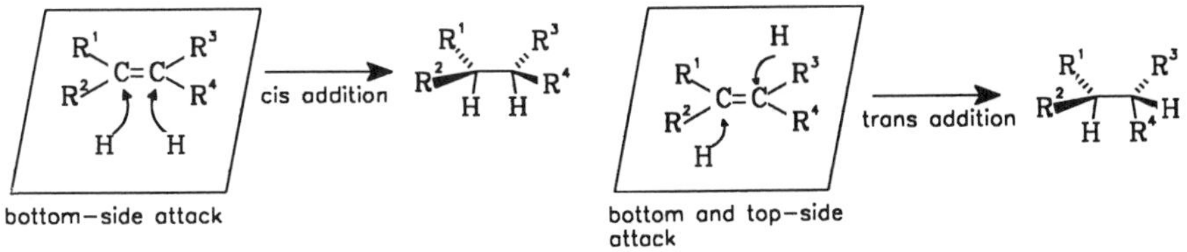

cis addition
bottom–side attack
trans addition
bottom and top–side attack

Scheme 1.

Moreover there is the stepwise addition with the possibility of bond rotation in the semi–hydrogenated state. A knowledge of the stereoselective course of hydrogenations of appropriate substrates may allow one to draw conclusions concerning the conformation of the chemisorbed molecule at the catalyst surface.

In diastereoselective hydrogenations, for example a knowledge of the absolute configuration of the newly formed stereogenic center allows one to draw conclusions concerning the conformation of the chemisorbed molecule at the catalyst surface, if the mode of hydrogen addition is fixed.

Thus, Horner [2] interpreted results of diastereoslective hydrogenations of α– and β– methyl–cinnamic acid derivatives (esters and amides) on the basis of the cis addition mechanism (scheme 2.).

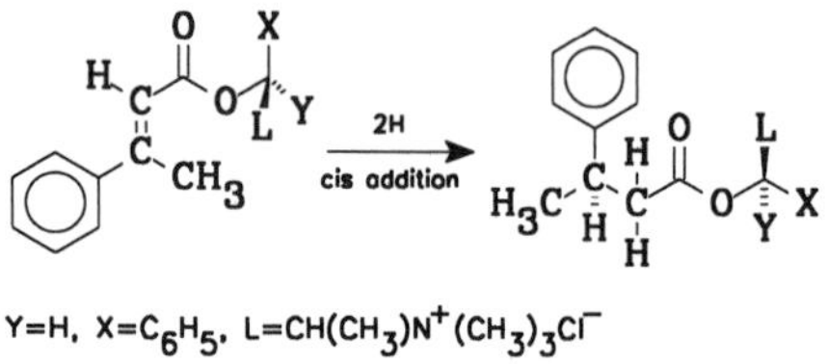

$Y=H$, $X=C_6H_5$, $L=CH(CH_3)N^+(CH_3)_3Cl^-$

Scheme 2.

RESULTS AND DISCUSSIONS

Recently we have reported [3–5] on the diastereoselective and enantioselective hydrogenation of Schiff–bases and iminium salts formed from chiral amines, for example (S)–proline, with prochiral ketones.

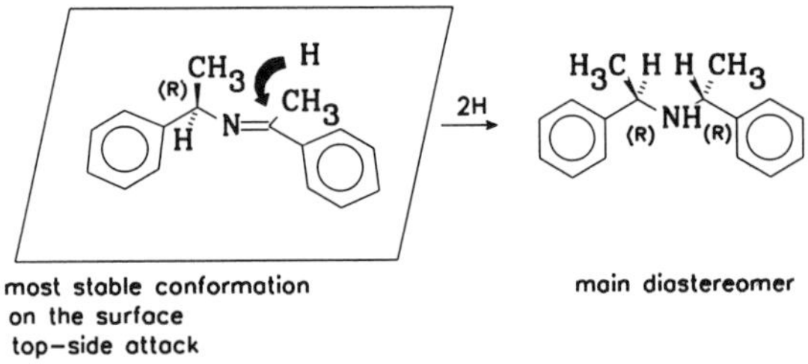

most stable conformation
on the surface
top–side attack

main diastereomer

Scheme 3. Diastereoselective hydrogenation of the Schiff–base of 1–phenyl–ethyl amine and acetophenone.

The observed configuration of the new stereogenic center together with an examination of the most favourable conformation of the chemisorbed substrates led us to the conclusion that hydrogen addititon occured from the top–side of the double bonds.

This mode of addition can possibly be explained on the basis of :

(i) the Horiuti–Polanyi mechanism on stepped surfaces, in holes, cavities and micropores or

(ii) so called "overlayer effect", i.e. an adsorbed layer of solvent molecules which can supply and stabilize activated hydrogen atoms.

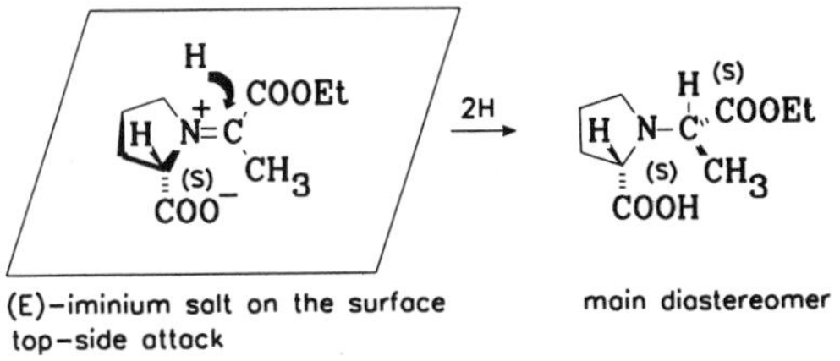

Scheme 4. Diastereoselective hydrogenation of the (E)–iminium salt formed from (S)–proline and ethyl pyruvate.

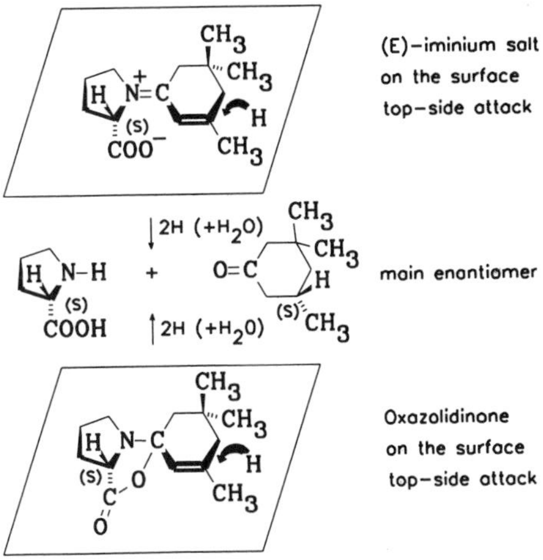

Scheme 5. Hydrogenation of isophorone in the presence of (S)–proline. Two possible intermediates.

An alternative explanation for the observed stereochemistry is that the minor, i.e. less stable, diastereomer of the catalyst–substrate complex is much more reactive than the major diastereomer and that it undergoes reaction via bottom–side attack. This is analogous to homogeneous rhodium–catalyzed asymmetric hydrogenations in which the product stereochemistry is controlled by the minor diastereomer of the catalyst–substrate complex.

Our findings, as did those of others [6] question the generality of the cis–addition mechanism of liquid phase catalytic hydrogenations and advocate a reevaluation of literature data.

REFERENCES

[1] M. Bartók: Stereochemistry of Heterogeneous Metal Catalysis.1985. John Wiley & Sons.
[2] L. Horner, H. Ziegler, H.D. Ruprecht, Liebigs. Ann. Chem. (1979) 341.
[3] A. Tungler, M. Ács, T. Máthé, E. Fogassy, Z. Bende, J. Petró, Appl. Cat., 17 (1985) 127.
[4] A. Tungler, T. Máthé, J. Petró, T. Tarnai, J. Mol. Cat., 61 (1990) 259.
[5] G. Tóth, A. Kovács, T. Tarnai, A. Tungler, Tetrahedron : Asymmetry Vol. 4.3 (1993) 331.
[6] F. van Rantwijk, A. van Vliet, H. van Bekkum, J. Mol. Catal., 9 (1980) 283.

HETEROGENEOUS ENANTIOSELECTIVE HYDROGENATION AND DIHYDROXYLATION OF CARBON CARBON DOUBLE BOND MEDIATED BY TRANSITION METAL ASYMMETRIC CATALYSTS

Dario Pini, Antonella Petri, Alberto Mastantuono, Piero Salvadori

Centro di Studio del C.N.R. per le Macromolecole Stereordinate ed
Otticamente Attive, Dipartimento di Chimica e Chimica Industriale,
Università di Pisa,
Via Risorgimento 35,
I–56126 Pisa, Italy.

INTRODUCTION

Transition metal complexes containing optically active ligands have proved to be among the most suitable homogeneous catalysts for the synthesis of optically active compounds. The main disadvantage of these homogeneous systems, which in general attain a high activity and enantioselectivity, arises from their difficult separation and recovery from the reaction medium. To overcome these problems, preventing the destruction of the catalytic species, transition metal complexes supported to insoluble matrices have been widely investigated. Nevertheless, by using such catalysts, very few examples of heterogeneous asymmetric reactions with significant enantioselectivity have been reported.

Two reactions that, in such sense, so far have furnished in heterogeneous phase a number of results comparable with those obtained in the homogeneous one, have been the hydrogenation and dihydroxylation of prochiral carbon–carbon double bonds.

We will deal with these two important reactions reviewing the most relevant results obtained up to now.

The heterogeneous enantioselective catalysts employed in the above reactions can be divided into two groups (scheme 1).

The type I includes catalytic systems obtained by combining transition metals with optically active low molecular weight modifiers, such as naturally occurring or easy to synthesize compounds. The type II is formed by asymmetric transition metal complexes, which are very active in the same homogeneous reactions, linked to organic (linear or cross–linked copolymer) or inorganic (silica gel or zeolites) polymeric backbones. In general the results attained were greatly affected by the structural characteristics of the insoluble matrices to which the catalytic sites were supported.

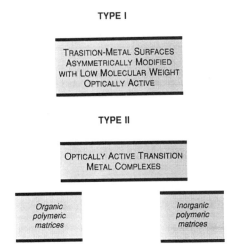

Scheme 1. Heterogeneous enantioselective catalysts.

HYDROGENATION

The heterogeneous catalytic enantioselective hydrogenation of olefinic bonds is a challenging area of the chemistry [1]. Up to date the efforts to make with a significant enantioselectivity this reaction on non–functionalized double bonds have not met with any success. In fact as far as to the best of our knowledge only the hydrogenation of α–ethylstyrene to (R)(–)–2–phenylbutane with a Merrifield resin supported chiral rhodium complex obtaining an optical purity of 1.5% was referred ! [2]

On the contrary several examples on the asymmetric hydrogenation of functionalized carbon carbon double bond were reported [1]. The different behaviour could be attributed to the fact that the substrate to be hydrogenated seems to demand the formation of a mixed ligand complex on the catalyst. The strength of the interactions between the components of this complex seems to depend on the presence of functional groups able to polarize the carbon carbon double bond and in addition to interact to some extent with the catalytic site. Practically cinnamic or acrylic derivatives (A,B) and acetamidocarboxylate olefinic compounds (C,D) have been the substrates of choice for the asymmetric hydrogenation (figure 1). Both types of catalysts reported in scheme I have been used.

Type I : Enantioselective Hydrogenation on Transition Metal Surfaces Asymmetrically Modified with Low Molecular Weight Optically Active Compounds.

Raney nickel (RNi), Pt and Pd have been the principal transition metals chirally modified or deposited on a chiral support. Many naturally occurring compounds (e.g. aminoacids, polypeptides, hydroxyacids, amines and aminoalcohols) or some enantiomerically pure synthetic molecules (e.g. α–phenyl or α–naphthyl–ethylamines and α–phenylpropyl amine) have been the most employed chiral modifiers for the metal surfaces [1].

In any case the hydrogenation of α,β–unsaturated acids (or salts) and esters by chirally modified metal surfaces gave rise to products with very low optical purity [3]. In table 1 and scheme 2 some significant examples are reported :

– optical yields higher than 10% have been obtained only in the case of methyl–(E)–α–phenylcinnamate (entry 3) and with its sodium salt (entry 4). Also the hydrogenation by modified Nickel Raney (TA–NaBr–MRNi) of various prochiral carboxylic acids salt gave products with low optical purity;

– very small degree of enantioselectivity (from 0.1 to 3.4%) has been registered [4,5] also in the catalytic hydrogenation of α–acetamidocinnamic acid on Pd(C) 5% in the presence of various optically active natural and synthetic amines.

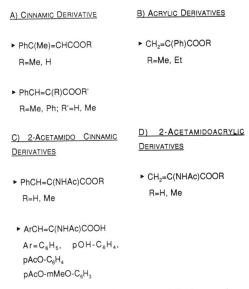

Figure 1. Substrates for asymmetric hydrogenation.

Table 1. Asymmetric hydrogenation of α,β–unsaturated acids or salts by chirally modified transition metal–surfaces.

entry	Substrate	Catalyst	O.Y.%
1	PhC(Me)=CHCOOH hydrocinchonine salt	Pt Adams	8-9
2	PhC(Me)=CHCOOH	Pd quinine, quinidine or cinchonidine salt / SiO$_2$	1.6-3.3
3	PhC(Me)=CHCOOH	Pd(C)/cinchonidine	30
4	(E)-PhCH=C(Ph)COO⁻Na⁺	TA-NaBr MRNI[a]	17

[a] TA NaBr MRNI: Modified Ni Raney with NaBr and tartaric acid

Type II : Enantioselective Hydrogenation with Optically Active Transition Metal Complexes Supported on Organic or Inorganic Polymeric Matrices.

Practically, so far, catalytic systems able to very well control the enantioselectivity have been obtained through heterogenization of some efficient chirally modified homogeneous rhodium(I) complexes, supporting them onto organic or inorganic polymers via phosphorus ligands (figure 2).

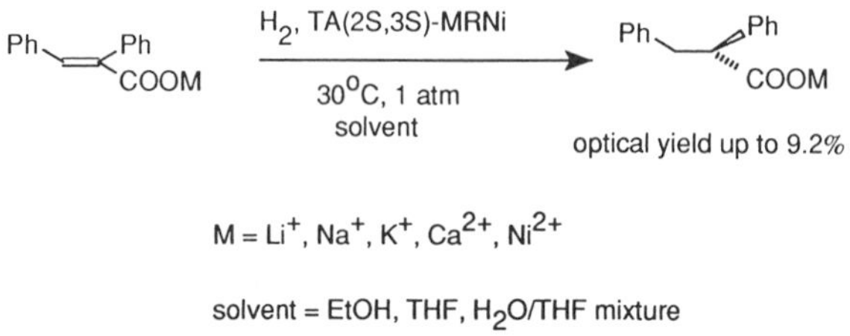

M = Li$^+$, Na$^+$, K$^+$, Ca^{2+}, Ni^{2+}

solvent = EtOH, THF, H$_2$O/THF mixture

Scheme 2.

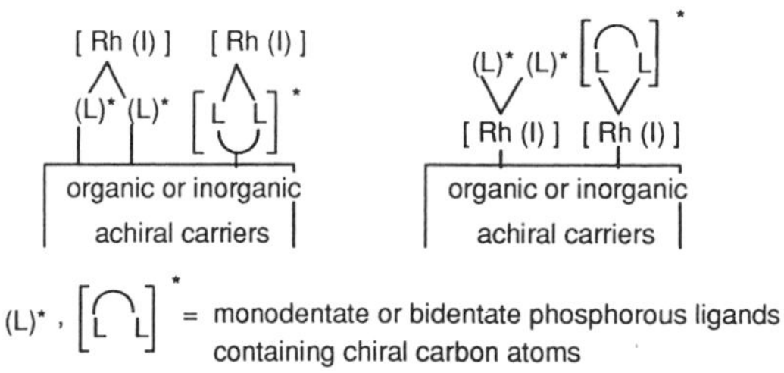

Figure 2. Heterogenization of some efficient chirally modified homogeneous rhodium(I) complexes on organic or inorganic polymers.

This catalyst can be obtained either (i) by supporting achiral rhodium(I) complexes on inherently chiral polymers or (ii) by bonding chiral rhodium(I) complexes to inherently achiral polymer supports. The chiral modifiers have been derived from whether optically active natural products (e.g. (+) tartaric acid, (−) menthol, 4−hydroxy−L−proline and glucose) or chiral synthetic molecules (e.g. 4−hydroxy−D−proline, (R,R)−3,4−dihydroxy− pyrrolidine and both enantiomers of 1−(α−naphthyloxy)−2−hydroxy−3−isopropyl− aminopropane). Few examples of the use of the (i) type catalyst have been referred : one of which has been the reduction of methyl−α−phtalimido acrylate to an alanine derivative (optical yield 28%) by an asymmetric catalyst prepared by reaction between cellulose modified with diphenylphosphino groups and achiral RhCl(PPh$_3$)$_3$ [6]. On the contrary, in the last ten years, a large number of examples with chiral rhodium complexes bound to an achiral organic or inorganic polymer has been reported, obtaining from good to excellent control of the enantioselectivity.

Chiral Rhodium(I) Catalysts on Organic Polymeric Matrices. Two types of insoluble optically active rhodium(I) catalysts, supported on organic polymeric matrices, have been performed. One has been obtained by copolymerization of a monomer carrying the bidentate phosphinic rhodium(I) ligand; the other has been prepared by anchorage of a cationic chiral rhodium(I) complex, as a whole, to a preformed cross–linked polymer, generally a commercial ion exchange resin.

In both cases, as a rule, two necessary requirements were observed for achieving the same high activity and enantioselectivity levels as those attained in the analogous homogeneous reaction : the polymer swelling in the reaction medium and the presence of a spacer between the chiral catalytic site and the polymeric main chain. Indeed only in this condition the chiral catalytic sites became accessible to the substrate. The most diffuse phosphorus bidentate chiral rhodium(I) ligands, suitable for the preparation of the copolymerization monomers, have been derived from the natural (R,R)–tartaric acid and (4S)–4–hydroxy–L–proline : 2–p–styryl–(4R,5R)–bis(diphenylphosphinomethyl)–1,3–di-oxolane, [(R,R)–DIOP derivative] and (2S,4S)– or (2R,4R)–N–methacryloyl–4–(diphenyl-phosphino)–2–diphenylphosphino)methyl]pyrrolidine, [(S,S)– or (R,R)–BPPM derivatives], respectively (scheme 3).

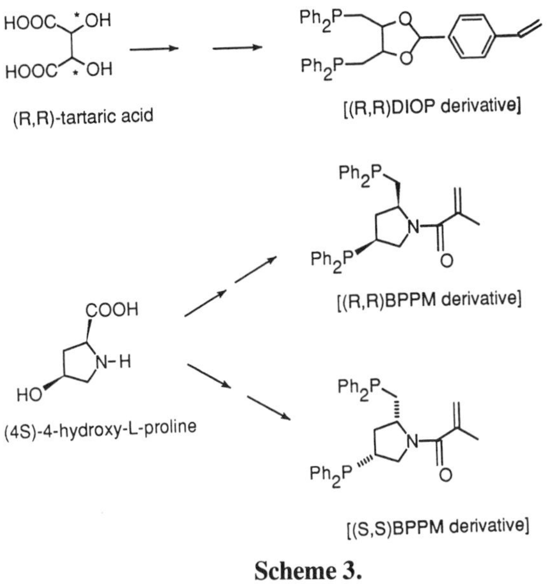

Scheme 3.

The cross–linked polymer–supported catalysts H1–Rh–(R,R)DIOP$_p$/HEMA and H1–Rh–(R,R)BPPM$_p$/HEMA or H1–Rh–(S,S)BPPM$_p$/HEMA (figure 3), containing the pendant alcohol group derived from the hydroxyethylmethacrylate unit (HEMA) have been prepared reacting [RhCl(C$_2$H$_4$)]$_2$ with the insoluble polymeric catalytic precursor. These catalysts have been employed in the asymmetric hydrogenation of α–acetamidoacrylic or cinnamic acids and α–phenylcinnamic acid [7–10] (scheme 4). As cross–linking agent ethylene dimethacrylate was added to obtain insoluble polymeric catalysts swelling in the polar solvents in which the substrates were soluble.

By using [7,8] H1–Rh–(R,R)DIOP$_p$/HEMA optically active products have been obtained in quantitative yield showing an optical purity up to 86%. The insoluble catalyst, isolated by filtration under an inert atmosphere, has been reused without significant loss of activity and enantioselectivity. It is therefore reasonable to assume that the rhodium leaching was prevented at least to a large extent.

It is noteworthy that a similar cross–linked DIOP–bearing polymer catalyst, devoid of pendant alcoholic groups and then unable to swell in protic solvent, was not suitable for catalyzing the asymmetric hydrogenation of polar substrates as α–acetamidoacrylic and cinnamic acids [2] .

Also with the polymer–bound catalysts H1–Rh–(R,R)BPPM$_p$/HEMA and H1–Rh–(S,S)BPPM$_p$/HEMA [9,10] high optical yields (up to 91%) have been achieved; in addition the heterogeneous system presented a high turnover. The values of the optical yield both in heterogeneous and in homogeneous phase, would suggest that the pendant alcohols do not interact with the catalytic active sites.

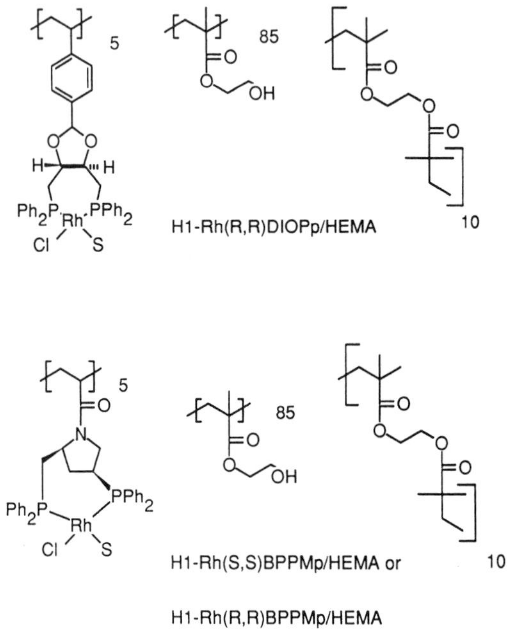

Figure 3. Cross–linked polymer–supported catalysts.

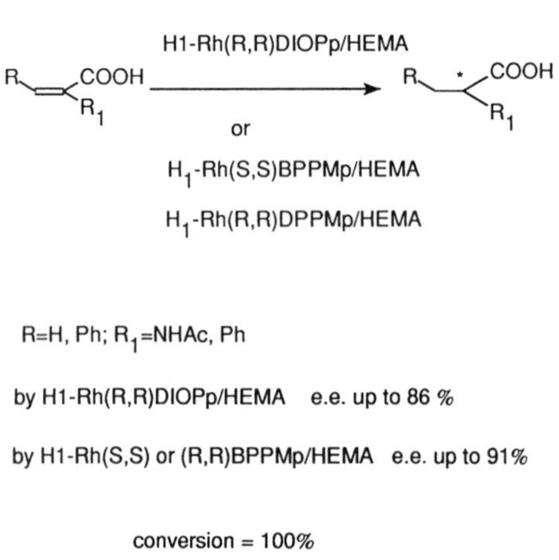

Scheme 4.

Nevertheless, to test if the introduction into the DIOP–bearing polymer matrix of an *ancillary* asymmetric center (i.e. an other optically active site which does not directly participate in the asymmetric hydrogenation) can affect the optical yields, a different cross–linked copolymer of 2–p–styryl–(4R,5R)–4,5–bis(diphenyl phosphinomethyl)–1,3–dioxolane with both the enantiomers of α–methylallyl alcohol (MAA) has been synthesized [11], H2–Rh(R,R)DIOP$_p$/(R)MAA and H2–Rh(R,R)DIOP$_p$/(S)MAA (figure 4). With these catalytic systems the hydrogenation of N–acetamidoacrylic acid has been examined [11] in alcoholic solvent and in THF. The polymers, swelling in both the solvents, were able to promote the asymmetric hydrogenation, even if the rates and the optical yields were strongly affected by the protic and solvating character of the solvent. Such results seemed to exclude some synergistic solvating effect of the chiral alcohol sites in the transition state of the enantioselective step.

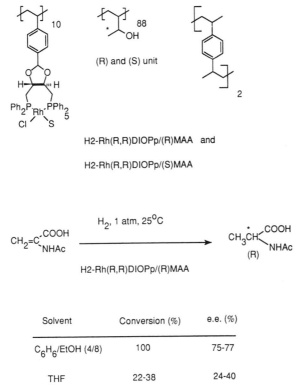

Figure 4. Cross–linked polymer with both enantiomers of α–methylallyl alcohol.

However a competition for the catalytic sites of ethanol, as solvent, with the polymer–bound alcohols was not to be excluded. In particular the low optical yields pointed out in THF could be attributed to the impossibility of some ancillary chiral alcohol sites, too much adjacent to the polymer backbone, to interact with the metal center because of their poor mobility. In order to overcome this drawback, some cross–linked polymers have been prepared [12], containing the alcoholic chiral center more spaced from the polymer backbone, owing to incorporation of (R,R)–, (S,S)– and (R,R)(S,S)–2,3–butanediol mono–acrylate units (figure 5). The results obtained in the hydrogenation at 20°C and 1 atm H$_2$ of α–acetamidoacrylic acid are compared both with those registered using the corresponding polymer catalysts without chiral alcohol units, H1–Rh–DIOP$_p$/HEMA and H1–Rh–BPPM$_p$/HEMA, and the soluble catalysts [DIOP–RhClS] and [BPPM–RhClS]. The most striking conclusions were that i) the polymer–bound alcohols participated at the catalyst site to provide an alcohol–like environment, ii) the ethanol as solvent swamped out the effect of the polymer–bound ancillary alcohol sites, iii) the pendant primary alcohols in the polymer matrix gave enantiomeric excesses higher than those provided by the more bulky chiral secondary alcohol. The last hypothesis seems to be strengthened by the results obtained in the hydrogenation of α–acetamido acrylic acid by using supported catalysts containing both a primary alcohol and a chiral secondary alcohol in the same pendant group [13] (figure 6).

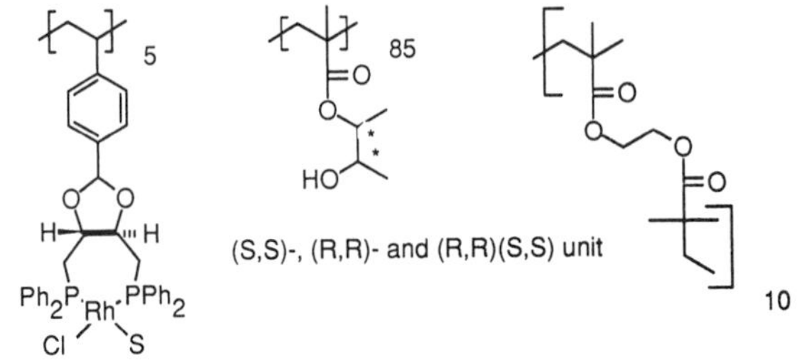

H3-Rh(R,R)DIOPp/(S,S)BDA; H3-Rh(R,R)DIOPp/(R,R)BDA

and H3-Rh(R,R)DIOPp/(R,R)(S,S)BDA

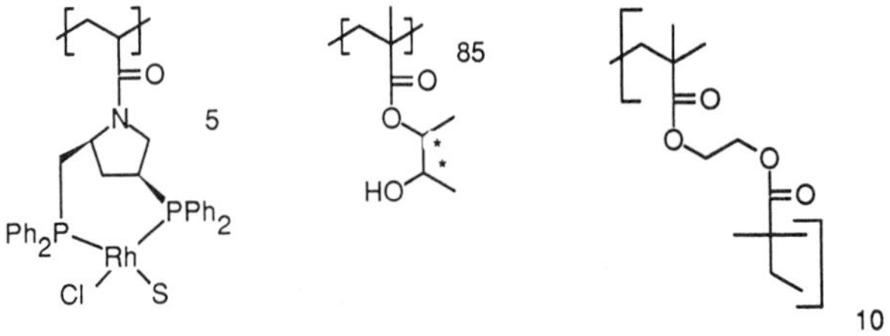

H3-Rh(R,R)BPPMp/(S,S)BDA; H3-Rh(R,R)BPPMp/(R,R)BDA

and H3-Rh(S,S)BPPMp/(R,R)(S,S)BDA

Figure 5. Cross–linked polymer containing alcoholic chiral center more spaced from the polymer blackbone.

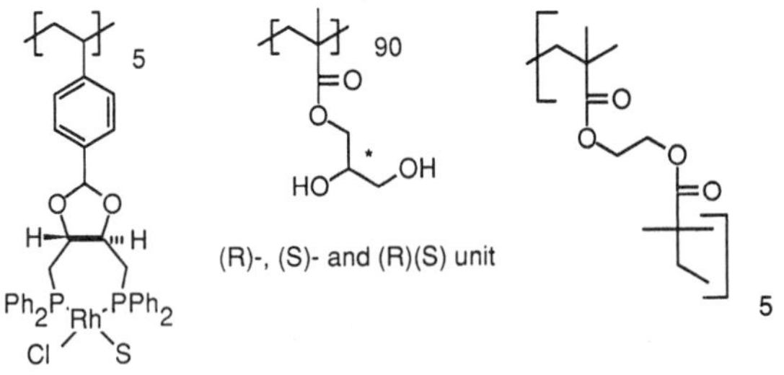

H3-Rh(S,S)DIOPp/(S)DHPMA; H3-Rh(S,S)DIOPp/(R)DHPMA

and H3-Rh(S,S)DIOPp/(R)(S)DHPMA

Figure 6. Supported catalyst containing both a primary alcohol and a chiral secondary alcohol in the same pendant group.

The other type of organic polymeric support is represented by cationic exchange resins.

The complex $[Rh(COD)Ph-\beta-glup]^+BF_4^-$, an excellent homogeneous chiral catalyst for the enantioselective hydrogenation of some α-acetamidoacrylic or cinnamic esters itself, has been used bound on commercially available cation exchange resin in the H^+ form such as the sulphonated styrene divinylbenzene (2%) copolymer (figure 7). This supported catalysts shows an enantioselectivity ca. 3% higher than the corresponding homogeneous ones (table 2); however in each case a lower activity has been registered, plausibly due to the poor substrate diffusion into the exchanger beads.

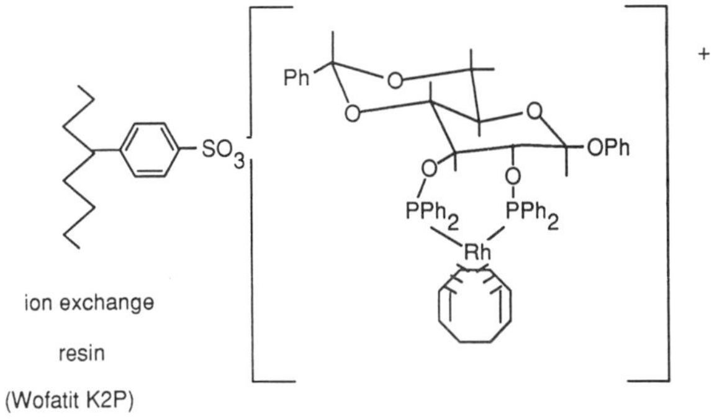

$[Rh(COD)Ph- \beta -glup]^+resin^-$

Figure 7. The complex $[Rh(COD)Ph-\beta-glup]^+BF_4^-$ bound on a commercial cation exchange resin in the H^+ form.

The recycling experiments revealed a very low metal leaching (ca. 5% after nine runs) and a moderate deactivation, caused by oxygen traces and/or ligand solvolysis and not by leaching, was registered after ten runs. As far as the enantiomeric excess trend was concerned, surprisingly it has been noted that after the first run it raised to a constant value ca. 1% higher than the initial one (in the case of the substrate A from 94 to 94.8%).

The lower enantioselectivity in the first run and in homogeneous experiments has been explained as resulting from the solvolytic cleavage of the complexed ligand on the benzylidene ring by the solution acidity (HBF_4) generated during the immobilization.

A confirmation of this hypothesis seemed to come from the results [14,15] with the heterogenized cationic rhodium phosphinite complex $[Rh(COD)Ph-\beta-glup-OH]^+resin^-$ (figure 8) obtained by loading the cationic resin in the neutral-form with the Rh-complex derived from the hydrolysis of $[Rh(COD)Ph-\beta-glup]^+BF_4^-$.

In table 3 a comparison of the hydrogenation of methyl α-acetamidocinnamate by using the heterogeneous catalysts $[Rh(COD)Ph-\beta-glup]^+resin^-$, $[Rh(COD)Ph-\beta-glup-OH]^+resin^-$ and the homogeneous ones $[Rh(COD)Ph-\beta-glup]^+BF_4^-$, $[Rh(COD)Ph-\beta-glup-OH]^+BF_4^-$ is reported. The above catalyst has been recycled many times without loss of neither activity nor enantioselectivity.

An other efficient cationic resin-supported chiral catalyst has been prepared [14] by anchoring to the sulphonated styrene-divinylbenzene copolymer the rhodium-complex formed from the inexpensive ligand 2,3-O,N-bis(diphenylphosphino)-1-(α-naphthyloxy)-2-hydroxy-3-isopropylaminopropane (PROPRAPHOS) readily available as both enantiomers (figure 9).

Table 2. Comparison between homogeneous and heterogeneous hydrogenation of some α–acetamidoacrylic and cinnamic methyl esters with [Rh(COD)Ph–β–glup] derived catalysts.

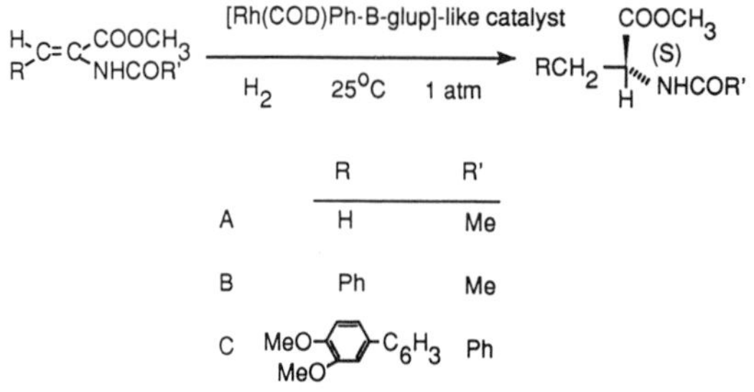

Subs	% e.e. (S)		Δ % e.e.	t/2 [b] (min)		$\upsilon_{het}/\upsilon_{hom}$ [c]
	homog.	heterog.		homog.	heterog.	
A	90.9± 0.5	94.3± 0.5	+3.4	0.9± 0.3	15± 2	0.06
B	91.1± 0.5	94.2± 0.5	+3.1	6.2 ±1.0	137± 21	0.05
C	82.5± 1.5	84.5± 1.5	+2.0	10.0± 2.0	180± 63	0.06

[a] The heterogeneous experiments were carried out with the same catalyst previously used in the homogeneous ones and then adsorbed onto cation resin, washed with methanol and separated from the solution. [b] Catalytic activity as time of half life of the substrate. [c] Ratio of hydrogenation rates.

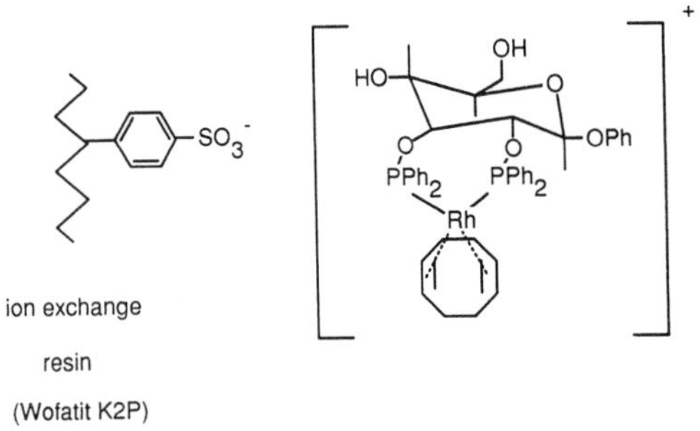

Figure 8. Heterogenized cationic rhodium phosphinite complex [Rh(COD)Ph–β–glup–OH][+]resin[−] obtained by loading the cationic resin in the neutral form with the rhodium complex.

Nevertheless, both in homogeneous and heterogeneous hydrogenation of α–acetamidoacrylic and cinnamic esters, this catalyst showed activity and enantioselectivity values much lower than those exhibited by [Rh(COD)Ph–β–glup]–like catalysts. When the catalyst was recycled, in general a loss of enantioselectivity has been obtained (from 79.9% in first run to 24.9% in ninth run), associated with a decrease of catalytic activity (t/2 increasing from 22 min in first run to 227 min in ninth run). With such cationic resin–supported catalysts the metal leaching was always very moderate, never overcoming 6%. Therefore, as already mentioned, the activity loss certainly was to be attributed to the chemical degradation of the catalytic complex.

Soluble form of the superacid exchange resin, Nafion® NR 50 [15] also has been used. The heterogeneous catalyst have been prepared by immobilizing on the resin the chiral rhodium complexes [(S,S)BDPP(NMe$_2$)$_4$RhNBD]$^+$BF$_4^-$ and [(S,S)Chiraphos (NMe$_2$)$_4$RhNBD)$^+$ BF$_4^-$ (figure 10). The results in the asymmetric hydrogenation of some dehydroaminoacid derivatives are shown in table 4 : a comparison with the result obtained in homogeneous phase is also given.

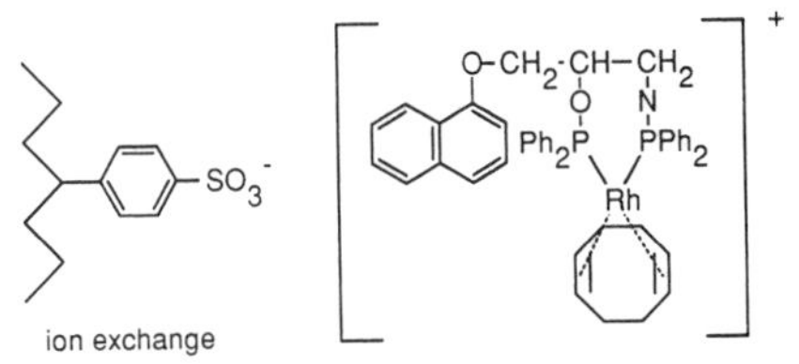

ion exchange

resin

(Wofatit K2P)

[Rh(COD)-PROPRAPHOS]$^+$resin$^-$

Figure 9. Cationic resin–supported chiral catalyst prepared from the rhodium complex [Rh(COD)–PROPRA– PHOS] and a sulphonated styrene–divinylbenzene copolymer.

Table 3. Homogeneous and heterogeneous hydrogenation of methyl α–acetamido– cinnamate with [Rh(COD)Ph–β–glup]– and [Rh(COD)Ph–β–glup–OH]– derived catalytic systems.

Complex	t/2 (min)	% e.e. (S)
[Rh(COD)Ph-β-glup]$^+$BF$_4^-$	6	91.1
[Rh(COD)Ph-β-glup-OH]$^+$BF$_4^-$	4	95.0
[Rh(COD)Ph-β-glup]$^+$resin$^-$	38	95.3
[Rh(COD)Ph-β-glup-OH]$^+$resin$^-$	17	95.5

[a]Reaction conditions: 1 mmol substrate, 0.01 mmol complex, methanol, 25°C, 1 atm H$_2$

Chiral Rhodium(I) Catalysts on Inorganic Polymeric Matrices. As far as the inorganic matrices is concerned very high enantioselectivity has been achieved also with some rhodium complexes heterogenized by anchoring them to silica support.

These complexes sometimes have shown higher stability and asymmetric efficiency than analogous homogeneous catalysts. Nevertheless one of the major problems presented by these heterogenized systems was the metal leaching during recycling.

By linking (EtO)$_3$Si(CH$_2$)$_n$P(menthyl)$_2$ to silica and loading it with [RhCl(C$_2$H$_4$)$_2$]$_2$ a series of chiral catalysts having a different length of the spacer (n=1, 3, 5) between the chiral active site and silica surface has been prepared [16] (figure 11).

These systems have been used in the hydrogenation of α–acetamidocinnamic acid , α– acetamidoacrylic acid and itaconic acid obtaining, in almost all cases, products with very high conversion and from good to excellent enantiomeric excess.

In the same time the asymmetric catalytic properties of the analogous homogeneous rhodium complexes have been studied.

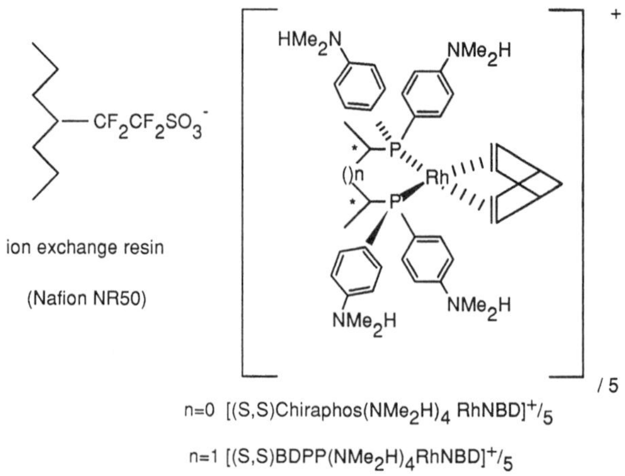

n=0 [(S,S)Chiraphos(NMe$_2$H)$_4$ RhNBD]$^+$/$_5$

n=1 [(S,S)BDPP(NMe$_2$H)$_4$RhNBD]$^+$/$_5$

Figure 10. Chiral complexes [(S,S)BDPP(NMe$_2$)$_4$RhNBD]$^+$BF$_4^-$ and [(S,S)Chiraphos(NMe$_2$)$_4$RhNBD]$^+$BF$_4^-$ immobilized on Nafion.

Table 4. Asymmetric hydrogenation with Nafion–supported or homogeneous complexes at atmosphere pressure.

substrate	Nafion-H supported[a]			Homogeneous[b]	
	Cycle	ee (%)	React. time (min)	ee (%)	React. time (min)
PhCH=C(COOMe)NHCOPh	1	50	15		
	2	41	40	57	5
	3	40	40		
	4	43	60		
PhCH=C(COOH)NHCOMe		65	21	91	9
PhCH=C(COOMe)NHCOMe		46	20	72	4

[a] 0.025 mmol [Rh(NBD)(BDPP)-(pNMe$_2$)$_4$]BF$_4$ on 110 m^2 Nafion. b 0.025 mmol [Rh(NBD)(BDPP)-(pNMe$_2$)$_4$]BF$_4$ in the presence of 0.1 ml HBF$_4$. Substrate/Rh=100; solvent: 10ml of MeOH, 20°C, 1 bar of H$_2$, conversion: 100%. The products are of R configuration.

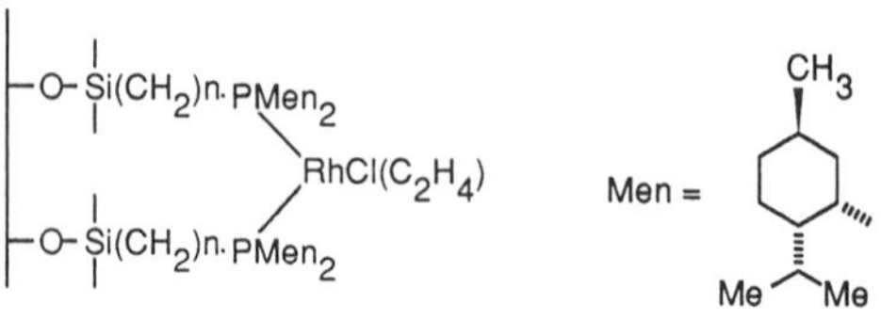

n = 1, 3, 5

Figure 11. Different lengths of spacer between chiral active site and support on chiral catalysts prepared by linking (EtO)$_3$Si(CH$_2$)$_n$P(menthyl)$_2$ to silica and loading with [RhCl(C$_2$H$_4$)$_2$]$_2$.

The catalytic activity and enantioselectivity of the supported catalysts depended not only on the structure of the unsaturated substrates but also on the length of the spacer and on the rhodium loading.

In the table 5 the most significant results obtained in the hydrogenation of α-acetamidocinnamic acid with these catalysts are reported, comparing them with those attained by soluble counterparts. It is worth mentioning that the enantioselectivity passes through a maximum and then decreases with increasing excess of ligand with respect to rhodium. A similar trend has been explained with the formation of various monomeric and dimeric ethylene–rhodium–phosphine species [16].

Table 5. Hydrogenation of α–acetamidocinnamic acid. Catalytic activity and enantioselectivity of heterogenized rhodium catalysts compared with the homogeneous ones.

Cat. n	Heterogeneous conditions				Homogeneous conditions			
	Rh content %	Time h	Conv. %	O.Y. %	P/Rh molar ratio	Time h	Conv %	O.Y. %
1	0.07	14.5	100	67 (R)	1	23	35	11 (R)
3	0.10	5.0	100	80 (R)	1	23	97	11 (R)
3	0.30	45.0	100	58 (R)	2	19.5	100	47 (R)
3	0.56	112.0	100	50 (R)	3	9.5	100	80 (R)
5	0.16	3.0	100	87 (R)	-	-	-	-

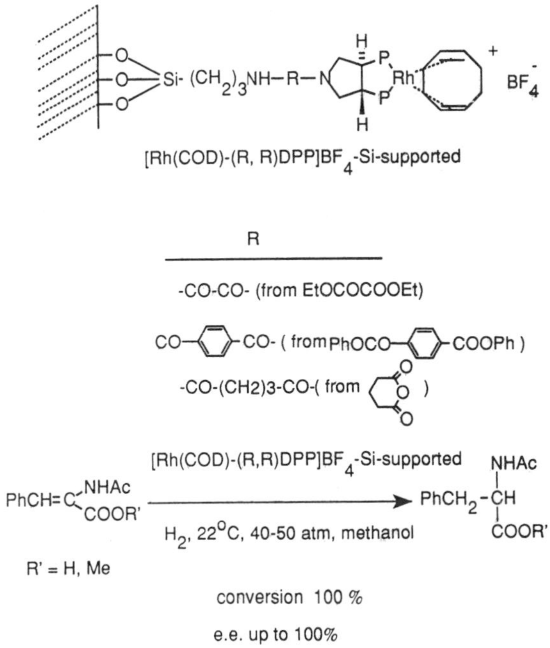

[Rh(COD)-(R, R)DPP]BF$_4$-Si-supported

R

-CO-CO- (from EtOCOCOOEt)

CO-⟨⟩-CO- (from PhOCO-⟨⟩-COOPh)

-CO-(CH2)3-CO-(from ⟨⟩)

[Rh(COD)-(R,R)DPP]BF$_4$-Si-supported

PhCH=C⟨NHAc⟩⟨COOR'⟩ $\xrightarrow{\text{H}_2, 22°C, 40-50 \text{ atm, methanol}}$ PhCH$_2$-CH⟨NHAc⟩⟨COOR'⟩

R' = H, Me

conversion 100 %

e.e. up to 100%

Scheme 5.

On recycling the catalysts, the total metal leaching was very high, coming after three runs up to 93% of the initial amount of rhodium, both in methanol and in THF. As a consequence, a marked loss in activity and in enantioselectivity has been registered.

An interesting example [17] of the use of chiral 1,2–diphosphines as ligands in the preparation of heterogenized rhodium complexes on silica gel characterized by a five-membered chelate ring, concerns the fixation of 3,4–(R,R)–bis(diphenylphosphino) pyrrolidine tetrafluoroborate, [Rh(COD)(R,R)DPP]$^+$BF$_4^-$, on the silica support. Some triethoxyaminopropyl–N–derivatives of [Rh(COD)(R,R)DPP]$^+$BF$_4^-$, having different length of links between the chiral moiety and triethoxysylil group, were reacted in different solvents with silica of various pore size (scheme 5).

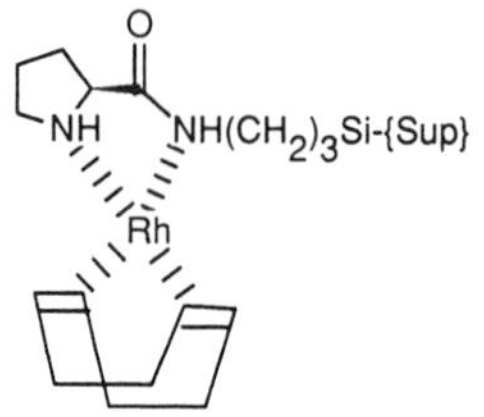

{Sup} = silica gel (Sil - 1) or

USY - zeolite (Zeol - 1)

Figure 12. Rh–complex with proline derivative anchored on supports.

Table 6. Data for catalytic hydrogenation of N–acylphenylalanine derivatives on anchored rhodium–proline complexes.

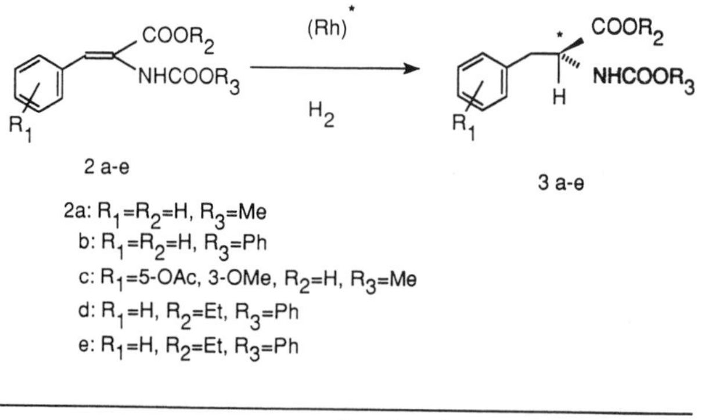

2 a-e

2a: R$_1$=R$_2$=H, R$_3$=Me
 b: R$_1$=R$_2$=H, R$_3$=Ph
 c: R$_1$=5-OAc, 3-OMe, R$_2$=H, R$_3$=Me
 d: R$_1$=H, R$_2$=Et, R$_3$=Ph
 e: R$_1$=H, R$_2$=Et, R$_3$=Ph

Compound	Conversion %[a] (Time h)[b]	Enantiomeric excess %		Confg. of **3**
		Sil-1	Zeol-1	
2a	100 (10)	88.0	97.9	R
2b	100 (14)	93.5	96.8	S
2c	100 (10)	-	94.0	R
2d	100 (10)	58.0	94.2	R
2e	100 (10)	92.2	99.0	S

[a] Measured by HPLC. [b] Using Zeol-**1**

The optical purity of the product varied very slightly with the length and the flexibility of the spacer; only in the case of methyl ester the silica pore size affected the enantioselectivity.

Also in this case the high enantiomeric excesses attained with the heterogeneous catalysts were comparable to those of their soluble counterpart (e.e. 99%) [18].

Rh–complexes with proline derivatives [19] (figure 12), when anchored on a modified USY zeolite which contains profuse supermicropores, produced a remarkable increase of enantioselectivity (>95%) in the hydrogenation of N–acyldehydrophenylalanine derivatives. The results obtained at 65°C and 5 atm of hydrogen are reported in table 6; in the same table also a comparison with the results obtained supporting compound **1** onto silica gel (Sil–**1**) is reported.

The phenylalanine derivatives were obtained with quantitative conversion and with high enantioselectivity and low dependence on temperature and hydrogen pressure.

From what has been described thus far it is evident that the catalytic heterogeneous asymmetric olefin hydrogenation shows preparative enantioselective validity only by using : i) organic or inorganic polymer–supported chiral rhodium complexes as catalysts; ii) α–acetamidocinnamic or acrylic acids and their esters as substrates. In these conditions it was possible to obtain in high enantiomeric excess, chiral α–aminoacids with both configurations.

DIHYDROXYLATION

Together with the oxidation to epoxides, the alkene dihydroxylation to 1,2 diols is one of the most interesting transformation of the olefinic bond. In particular, when asymmetrically accomplished on a prochiral substrate, this oxidative reaction is of great importance to obtain two vicinal hydroxyl groups stereospecifically embedded in compounds of defined chirality; this chiral diols show very frequently biological activity [20], or can be widely employed as chiral auxiliaries in asymmetric syntheses [21].

For better understanding the nature of the dihydroxylation reaction and how it has been improved during the recent years, until to heterogeneous asymmetric catalytic version, it seems useful to review its historical background.

Osmium tetroxide (OsO_4) has been the most reliable reagent for the cis dihydroxylation of the olefins [22]. Even though initially it has been used prevailingly in stoichiometric amount, the possibility of make the process in a catalytic way, in the presence of a secondary oxidant, has represented a remarkable success because of the high cost and toxicity of this metal oxide. As secondary oxidants, many compounds were used in the reaction, e.g. sodium or potassium chlorate, hydrogen peroxide, t–butyl hydroperoxide, N–methylmorpholine–N–oxide (NMO), potassium ferricyanide [23], etc..

It was also observed that by addition of a nucleophilic ligand, such as pyridine and other tertiary amines and diamines, the reaction rate was markedly enhanced [24].

A further improvement, concerning a more practical use of this catalytic process, was reached employing OsO_4 coordinated to an amine moiety linked to a polymeric backbone, obtaining an heterogeneous dihydroxylation system [25].

For this purpose, commercial cross–linked poly–4–vinylpyridine or macromolecular basic ligands synthesized by quaternization of tertiary amines (e.g. 1,4–diaza–bicyclo[2.2.2]octane, hexamethylenetetramine, N,N,N',N'–tetramethylenediamine) with chloromethylated styrene–divinylbenzene insoluble copolymers were used. These polymeric systems can offer the well known advantages of the heterogeneous catalysts and in addition, in the case of a toxic reagent such as OsO_4, the safe handling.

In the last years, most work was done on the asymmetric version of the OsO_4 mediated reaction on prochiral substrates, using amines and diamines as chiral auxiliaries and a review recently appeared dealing with the last advances obtained in this field [26].

Even though several ligands are used in stoichiometric amount giving high optical yields, on the contrary, in the catalytic version of this reaction to date significative results were obtained only with derivatives of two Cinchona alkaloids , quinine and quinidine.

As far as the homogeneous asymmetric catalytic dihydroxylation is concerned, the first examples appeared in 1988 [27,28]; afterwards the major development was given by Sharpless and co–workers [29]. In the original procedures, dihydroquinine and dihydroquinidine esters were used as chiral ligands in the presence of catalytic amount of OsO_4 and NMO as secondary oxidant in a mixture of water and acetone (figure 13).

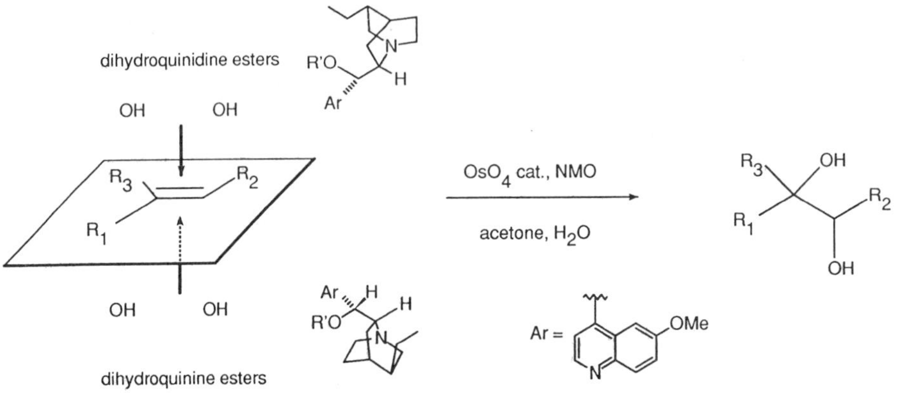

Figure 13. Dihydroquinine and dihydroquinidine esters used as chiral ligands in presence of catalytic amount of OsO_4 and NMO as secondary oxidant in a mixture of acetone and water.

By the "quasi" enantiomer relationship between the two chiral auxiliaries both antipodes of the oxidation product were obtained from the same olefin, practically with the same chemical and optical yields.

An important experimental feature was the very slow addition (5–24 hours) of the olefin to the reaction mixture, which is necessary to obtain high e.e. : the reason of this experimental request was explained on the basis of the reaction mechanism hypothesis proposed by Sharpless [30]. The employed olefins contained aromatic and/or aliphatic groups, but the e.e. obtained with aromatic olefins were higher than those obtained with aliphatic ones.

During the past four years, the reaction in the homogeneous phase has been improved steadily, introducing some changes in the experimental procedure and in the nature of the chiral ligands.

The use of $K_3Fe(CN)_6$ instead of NMO as secondary oxidant, and t–BuOH/H$_2$O mixture instead of acetone/H$_2$O solution as reaction solvent, has allowed to obtain high enantioselectivity without the slow addition of olefin indispensable for the NMO system.

Also the nature of the chiral ligands was changed : two new families of alkaloid derivatives were used, the 9–O–(9'–phenanthryl)–dihydroquinidine and 9–O–(4'–methyl–2'–quinolyl)–dihydroquinidine [31] and, more recently, the C$_2$–symmetric phtalazine derivatives [29]. In these new experimental conditions the asymmetric cis dihydroxylation has been extended to a large class of olefinic substrates.

In table 7 a comparison is reported between the e.e. obtained in the dihydroxylation of some olefins in the different experimental conditions till now used.

Table 7. Enantiomeric excesses (%) of the diols resulting from gradually improved homogeneous asymmetric catalytic dihydroxylation (some significant examples).

OLEFINS	Monophase system NMO alkaloid ester derivatives	Biphase system $K_3Fe(CN)_6$ alkaloid ester derivatives	Biphase system $K_3Fe(CN)_6$ alkaloid ether derivatives	Biphase system $K_3Fe(CN)_6$ alkaloid phtalazine derivatives
styrene	60	74	87	97
1-decene	40	45	74	84
trans-stilbene	80	99	99	99.5
3-hexene	70	79	95	97
2-methyl-1-phenyl-1-propene	53	74	84	93

[a] In all cases the yields of diols were from 75 to 95%.

As shown in the table, together with high chemical yields, for many substrates, quasi complete enantioselectivity (≥ 99) was reached.

For further improve the convenience and economy of the process, an highly attractive approach is represented by the asymmetric heterogeneous catalytic dihydroxylation of the olefins employing insoluble polymer–bound chiral derivatives. To date only few notes have appeared dealing with this procedure. For this reaction, the heterogenized catalysts were obtained in all cases by linking the chiral catalytic sites on organic polymeric matrices (Type II, scheme 1).

Insoluble copolymers between some quinine derivatives and acrylonitrile were used in the preparation of OsO_4 chiral adducts for the catalytic asymmetric oxidation of olefins in the heterogeneous phase [32] (figure 14).

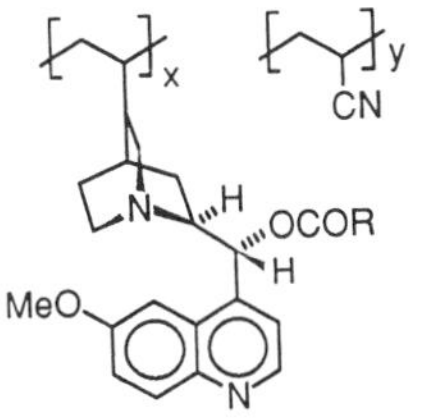

$1\ R = CH_3$

$2\ R = pCl\text{-}C_6H_4$

$3\ R = (mCH_3O)_2\text{-}C_6H_3$

Figure 14. Insoluble copolymer of quinine derivatives and acrylonitrile used in the preparation of OsO_4 chiral adducts for the catalytic asymmetric oxidation of olefins in the heterogeneous phase.

By using the copolymer containing 10% by mole of quinine acetate units, optically active diols have been obtained from several olefinic substrates in good chemical yields (up to 76%) and e.e. up to 30% (table 8).

Trans–stilbene has been used as model substrate in the dihydroxylation runs (table 9) employing copolymers with different level of alkaloid incorporation (see figure 14). The e.e. obtained (up to 46%) were lower than those provided by the homogeneous process and only slightly influenced by the polymeric alkaloid nature. In addition, high chemical yields in 1,2 diol were obtained only using copolymers with low alkaloid content (below 15% by mole).

Table 8. Heterogeneous catalytic asymmetric dihydroxylation of various olefins.

L^*P = insoluble polymeric ligand

R$_1$	R$_2$	Yield (%)	E.E. (%)	Confg.
C$_6$H$_5$	H	65	22	(S)
mCH$_3$-C$_6$H$_4$	H	70	28	(S)
C$_6$H$_5$	C$_6$H$_5$	75	24	(S,S)
nC$_5$H$_{11}$	H	10	6	(S)
COOCH$_3$	COOCH$_3$	76	30	(R,R)
C$_6$H$_5$	CH$_2$OAc	70	30	(R,R)

[a]10% by mol of alkaloid incorporation; T = 40 °C, 16 hrs; molar ratio olefin/OsO$_4$/polymer **1** = 1/0.05/0.05.

The OsO$_4$/polymeric alkaloid complex could be simply and quantitatively recovered at the end of the reaction and then reused 4–5 times without significant losses of activity or enantioselectivity. For further use, it was necessary to add a little amount of osmium tetroxide (about 0.1%) because of the activity and enantioselectivity reduction, probably due to osmium leaching.

Another paper, almost contemporary, about the heterogeneous asymmetric dihydroxylation of the *trans*–stilbene appeared [33], employing polymer bound Cinchona alkaloid derivatives (figure 15). With a polymeric support containing no spaced alkaloid units, 1,2–diol has not been obtained; on the contrary, the introduction of a spacer increases very much the activity (chemical yields up to 96%) and enantioselectivity (e.e. up to 87%).

Dihydroquinidine on a polystyrene support (figure 16) was also reported for the osmium catalyzed asymmetric dihydroxylation of alkenes [34]. The most effective polymeric ligand appeared to be that with 10% by mole of Cinchona alkaloid content. The enantioselectivity obtained was good (up to 69% for E–2–octene), but in some cases it was necessary to add the olefin slowly to the reaction mixtures. It is to note that in this paper, it is not clear if the reaction takes place in completely heterogeneous phase, since in an end-note it was referred that "*the reaction mixture gradually changed from a heterogeneous to a homogeneous solution in 18–20 h*".

Another alkaloidic monomer, containing the spaced chiral catalytic sites, was copolymerized with both acrylonitrile, obtaining a linear copolymer [35], and with styrene/divinylbenzene, yielding a new cross–linked polymeric precursor [36] (figure 17). In table 10 the dihydroxylation runs using the linear copolymer are depicted : the best results were obtained with *trans*–stilbene and $K_3Fe(CN)_6$ as secondary oxidant.

Table 9. Heterogeneous catalytic asymmetric dihydroxylation of *trans*–stilbene on various OsO_4/copolymers.

Copolymer[b]	T (°C)	Yield (%)	E.E. (%)
1 (25)	25	0	-
1 (10)	25	71	23
1 (25)	0	73	38
1 (10)	-15	70	46
1 (3)	25	65	19
2 (15)	25	70	15
2 (4)	25	73	14
2 (4)	0	71	27
3 (20)	25	0	-
3 (7)	25	83	30
3 (7)	0	86	45

[a]21 hrs; molar ratio olefin/OsO₄/polymeric alkaloid = 1/0.001/0.029. [b]In parentheses the alkaloid content (% by mol) is indicated.

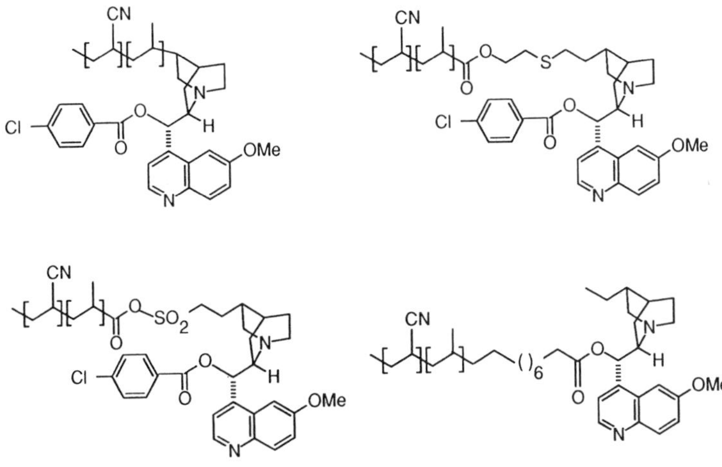

Figure 15. Polymer bound Cinchona alkaloid derivatives for dihydroxylation of *trans*–stilbene.

With the cross–linked copolymer optical yields comparable to those achieved in the homogeneous phase by employing Cinchona alkaloids p–chlorobenzoate derivatives, were attained not only with *trans*–stilbene but also with other olefins (table 11).

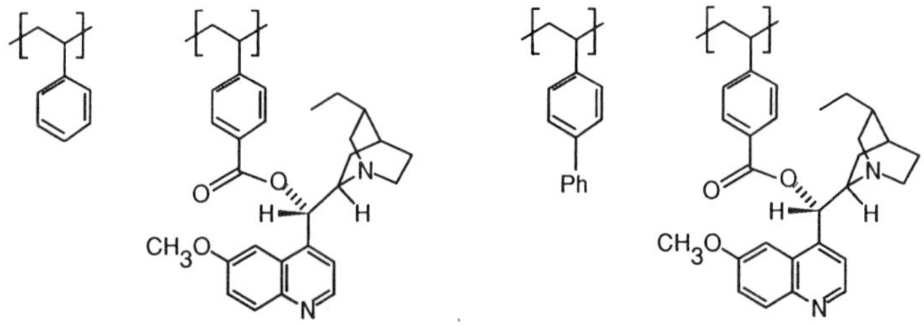

Figure 16. Dihydroquinidine on polystyrene support for the osmium catalyzed asymmetric dihydroxylation of alkenes.

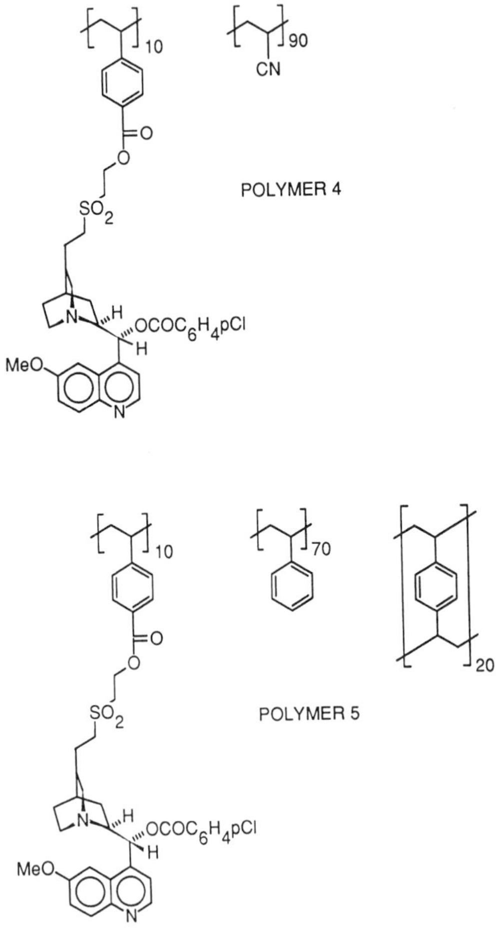

POLYMER 4

POLYMER 5

Figure 17. Alkaloidic monomer copolymerized with both acrylonitrile, obtaining a linear copolymer, and with styrene/divinylbenzene yielding a new cross–linked polymeric precursor.

The reactivity of the polymeric system is strictly dependent on the reaction conditions : by using potassium ferricyanide as secondary oxidant, the catalytic reaction is completely inhibited.

Table 10. Heterogeneous catalytic asymmetric dihydroxylation of olefins with OsO_4/polymer **4** system (see fig. 17).

Olefin	Secondary oxidant	Yield (%)	E.E. (%)
trans-Stilbene	NMO[b]	100	55
"	$K_3Fe(CN)_6$[c]	100	80
trans-β-Methylstyrene	"	70	36
Ethyl-4-nitrocinnamate	"	65	30
Styrene	"	90	41
1-Phenyl-1-cyclohexene	"	83	41

[a] T = 25°C, t = 20 hrs; [b] Acetone/H_2O 10/1 as solvent; molar ratio olefin/OsO_4/polymer **4** =1/0.001/0.029; [c] '$BuOH$/H_2O 1/1 as solvent; molar ratio olefin/OsO_4/ polymer **4**= 1/0.01/0.25.

Table 11. Heterogeneous catalytic asymmetric dihydroxylation of olefins with OsO_4/polymers **5** system (see fig. 17).

Olefin	Secondary Oxidant	T (°C)	t (hrs)	Yield (%)	E.E. (%)
trans-Stilbene	$K_3Fe(CN)_6$[a]	25	24	0	–
"	NMO[b]	25	6	81	69
"	NMO	0	7	85	87 (78)[c]
Styrene	$K_3Fe(CN)_6$	25	24	0	-
"	NMO	0	5	72	54 (56)
trans-β-Methylstyrene	NMO	0	6	69	60 (65)
1-Phenyl-1-cyclohexene	NMO	0	24	46	5 (8)

[a] '$BuOH$/H_2O = 1/1 as solvent: molar ratio olefin/OsO_4/polymer **5** = 1/0.01/0.25; [b] Acetone/H_2O = 10/1 as solvent: molar ratio olefin/OsO_4/polymer **5** = 1/0.001/0.029; [c] In parentheses the e.e. values obtained in the homogeneous phase with dihydroquinidine 9-O-p.chlorobenzoate as chiral catalytic precursor are reported.

This result can be reasonably ascribed to the fact that the polymer collapses in the protic polar solvent mixture t–$BuOH$:H_2O 1:1 used with potassium ferricyanide, preventing the substrate penetration. On the contrary, when NMO is used as secondary oxidant, the solvent is a 10:1 acetone:H_2O mixture and the reaction proceeds. All the results are obtained in a short reaction time and without the slow addition of the olefin; a temperature lowering led to an improvement of the e.e. of the diols.

To sum up the catalytic asymmetric dihydroxylation of the olefins is also a suitable reaction to be transferred from the homogeneous phase to the heterogeneous one, reaching the same levels of activity and enantioselectivity. Nevertheless, the drawback of the leaching of the osmium tetroxide, the more expensive and toxic reaction component, has not been completely overcome.

REFERENCES

[1]   Blaser H.U., *Tetrahedron Asymmetry,* **1991**, *2*, 843 and literature cited therein.
[2]   Dumont W., Poulin J.C., Dang, T.P., Kagan, H.B., *J. Am. Chem. Soc.,* **1973**, *95*, 8295.
[3]   Bartok M., Wittmann G., Bartok G.B., Gondos G., *J. Organomet. Chem.,* **1990**, *384*, 385.
[4]   Yoshida T., Harada K., *Bull. Soc. Chem. Jpn.,* **1971**, *44*, 1062.
[5]   Pracejus H., Bursian M., *East Ger. Pat.* 92 031.
[6]   Takaishi N., Imai H., Bertelo C.A., Stille J.K., *J. Am. Chem. Soc.,* **1976**, *98*, 5400.
[7]   Takaishi N., Imai H., Bertelo C.A., Stille J.K., *J. Am. Chem. Soc.,* **1978**, *100*, 264.
[8]   Baker G.L., Fritschel S.J., Stille J.K., *ACS Polymer Preprints,* **1981**, *22*, 155.
[9]   Baker G.L., Fritschel S.J., Stille J.R.,Stille J.K., *J. Org. Chem.,* **1981**, *46*, 2954.
[10] Masuda T., Stille J.K., *J. Am. Chem. Soc.,* **1978**, *100*, 268.
[11] Baker G.L., Fritschel S.J., Stille J.K., *J. Org. Chem.,* **1981**, *46*, 2960.
[12] Deschenaux R., Stille J.K., *J. Org. Chem.,* **1985**, *50*, 2299.
[13] Selke R., Haupke K., Krause W., *J. Mol. Catal.,* **1989**, *56*, 315.
[14] Selke R., *J. Mol. Catal.,* **1986**, *37*, 227.
[15] Tóth I., Hanson E.B., *J. Mol. Catal.,* **1992**, *71*, 365.
[16] Kinting A., Krause H., Capka M., *J. Mol. Catal.,* **1985**, *33*, 215.
[17] Nagel U., Kinzel E., *J. Chem. Soc., Chem. Commun.,* **1986**,1098.
[18] Nagel U., *Angew. Chem. Int. Ed. Engl.,* **1984**, *23*, 435.
[19] Corma A., Iglesias M., del Pino C., Sanchez F., *J. Chem. Soc., Chem. Commun,* **1991**, 253.
[20] a) Muraoka O., Fujiara N., Tanabe G., Momose T., *Tetrahedron Lett.,* **1991**, *2*, 357.
       b) Brosa C., Pracaula R., Puig R., Ventura M., *Tetrahedron Lett.,* **1992**, *33*, 7057.
       c) Wang Z.M., Zhang X.L., Sharpless K.B., Sinha S.C., Sinha–Bagchi A., Keinan E., *Tetrahedron Lett.,* **1992**, *33*, 6407.
       d) Turpin J.A., Weigel L.O., *Tetrahedron Lett.,* **1992**, *33*, 6563.
       e) Keinan S.C., Sinha S.C., Sinha–Bagchi A., Wang Z.M., Zhang X.L., Sharpless K.B., *Tetrahedron Lett.,* **1992**, *33*, 6411.
       f) Gurjar M.K., Mainkar *Tetrahedron Asymmetry,* **1992**, *3*, 21.
[21] Blaser H.U., *Chem. Rev.,* **1992**, *92*, 935.
[22] a) Schröeder M., *Chem. Rev.,* **1980**, *80*, 187.
       b) Courtney J.L., in *Organic Syntheses by Oxidation with Metal Compounds* (Eds. Meijs W.J. and De Jonge C.R.H.I.) Plenum Press, New York, 1986, p.445.
[23] Minanto M., Yamamoto K., Tsuji J., *J. Org. Chem.,* **1990**, 55, 766.
[24] Crieege, R. *Justus Liebigs Ann. Chem.* **1936**, *522*, 75.
[25] Cainelli G., Contento M., Manescalchi F., Plessi L., *Synthesis,* **1989**, 45.
[26] Lohray B.B., *Tetrahedron Asymmetry,* **1992**, *3*, 1317.
[27] Nardi A., *Tesi di Laurea*, Università di Pisa, Italy, April 1988.
[28] Jacobsen E.N., Markò I., Mungall W.S., Schröder G., Sharpless K.B., *J. Am. Chem. Soc.,* **1988**, *110*, 1968.
[29] Sharpless K.B., Amberg W., Bennani Y.L, Crispini G.A., Hartung J., Jeong K.S., Kwong H.L Morikawa K., Whang Z.M., Xu D., Zhang X.L., *J. Org. Chem.,* **1992**, *57*, 2768 and literature cited therein.
[30] Wai J.S.M., Markò I., Svendsen J.S., Finn M.G., Jacobsen E.N., Sharpless K.B., *J. Am. Chem. Soc.,* **1989**, *111*, 1123.
[31] Sharpless K.B., Amber W., Beller M., Chen H., Hartung J., Kawanami Y., Lubben D., Manoury E., Ogini Y., Shibata T., Ukita T., *J. Org. Chem.,* **1991**, *56*, 4585.
[32] a) Pini D., Rosini C., Nardi A., Salvadori P., *Fifth IUPAC Int. Symposium "OMCOS V"*, Abstr. PS–67, Florence (Italy), **1989.**
       b) Pini D., Petri A., Nardi A., Rosini C., Salvadori P., *Tetrahedron Lett.,***1991**, *38*, 5175.
[33] Kim M.B., Sharpless K.B., *Tetrahedron Lett.,* **1990**, *31*, 3003.
[34] Lohray B., Thomas A., Chittari P., Ahuja J.R., Dhal P.K., *Tetrahedron Lett.,***1992**, *33*, 5453.
[35] Pini D., *Seminars in Organic Synthesis*, XVIII Summer School "A. Corbella", June 14–18, Gargnano (Bs) Italy, **1993**, pg. 187–216.
[36] Pini D., Petri A., Salvadori P., *Tetrahedron Asymmetry*, **1993**, *4*, 2351.

SECTION IV

MORE ADVANCED REACTIONS

CHIRAL DIOXO-MOLYBDENUM COMPLEXES ANCHORED TO MODIFIED USY-ZEOLITES. APPLICATION TO SELECTIVE EPOXIDATION OF OLEFINS

A. Corma[1], M. Iglesias[2], J.R. Obispo[3] and F. Sánchez[3]

[1] Instituto de Tecnología Química, CSIC-UPV
Camino Vera s/n
E-46071 Valencia, Spain
[2] Instituto de Ciencia de Materiales, Sede D
Serrano 113
E-28006 Madrid, Spain
[3] Instituto de Química Orgánica
J. de la Cierva, 3
E-28006 Madrid, Spain

ABSTRACT

Chiral dioxo acetylacetonate Mo(VI)-complexes were synthesized by reactions of $MoO_2(acac)_2$ with a series of ligands derived from proline and 4-hydroxyproline. The structures of these complexes were elucidated by analytical and spectroscopic measurements. The complexes with $Si(OEt)_3$ groups were anchored to modified USY-zeolite and their catalytic activities tested in epoxidation reactions of alkenes with TBHP, as oxygen source. The heterogenized catalysts have shown a significant rate enhancing and a shape selectivity when compared with the corresponding homogeneous complexes.

INTRODUCTION

Chiral unfunctionalized aliphatic oxiranes are reactive chirons for the synthesis of optically active natural, pharmaceutical, and other intermediates products [1,2]. Non-racemic aliphatic oxiranes are accessible by a number of fundamental enantioselective strategies : chiral-carbon pool transformation, enzymatic and non enzymatic asymmetric epoxidation, kinetic resolution, and chromatographic resolution. While enantiomerically highly enriched α-(hydroxyalkyl) oxiranes can be obtained by asymmetric Katsuki-Sharpless [1] epoxidation of allylic alcohols with *tert*-butyl hydroperoxide, titanium (IV), and dialkyl tartrates, no such versatile procedure is yet available for the peroxometal-mediated epoxidation of unfunctionalized aliphatic alkene and still represents a particular challenge because of the lack of pendant groups favoring stereocontrol by conformational rigidity via auxiliary interactions. The preparation of chiral oxiranes by transition metal–catalyzed epoxidations have been reviewed by Jorgensen [2]. Only moderate enantiomeric yields were observed in Mo(VI)L–catalyzed asymmetric epoxidation of alkenes with *tert*–butyl hydroperoxide, where L= N–ethyl ephedrine, N–methyl prolinol and tartrate esters [2]. The aminoacids form the biggest pool of compounds and are used more and more frequently as auxiliary agents in asymmetric synthesis, and they are of interest because of their biological importance and the variety of metal coordination modes they can display, acting as ligands.

Optically active transition metal compounds can be anchored on surface rich supports [2], so that they do not become dissolved during the catalysis. Frequently, these immo–bilized compounds give an optical induction similar to their soluble counterparts. Heterogenized homogeneous catalysts combine the properties of heterogeneous and homogeneous catalysts, although a heterogenized homogeneous catalyst may lose its activity and enantioselectivity owing to deactivation and/or metal leaching [3].

In this paper we describe the synthesis and characterization of a series of chiral $MoO_2(acac)L$ where **L=1–5** (scheme 1) prepared from (*L*)–prolinol and (*L*)–*trans*–4–hydroxyproline.

Those compounds with a $Si(OEt)_3$ group were heterogenized by anchoring into silica and USY–zeolites and we report these novel heterogeneous catalytic systems for the shape selective epoxidation of alkenes.

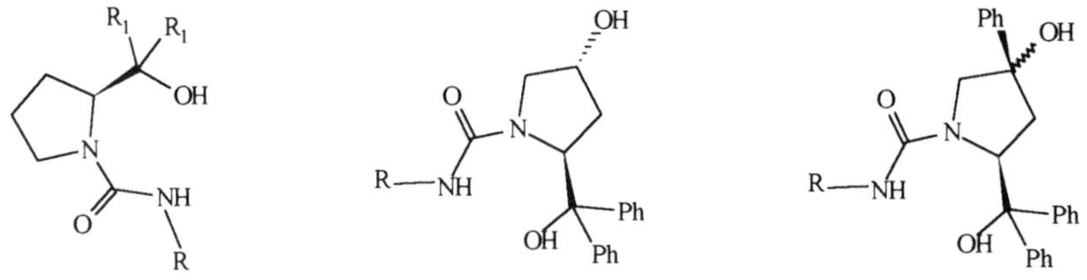

1,2a-b   3a-b   4,5a-b

1: R_1 = H; 2: R_1 = Ph   4: 2S,4R; 5: 2S,4S

a: R = t-Bu; b: R = $(CH_2)_3Si(OEt)_3$

Scheme 1.

EXPERIMENTAL SECTION

Reactions were carried out under oxygen–free nitrogen by Schlenk–tube techniques. C, H and N analysis were carried out by the Analytical Department of the Institute of Organic Chemistry (C.S.I.C.) with a Heraeus apparatus. The metal contents were determined by atomic absorption measures in a Unicam (SP9) Philips apparatus. Infrared spectra were recorded with a Nicolet XR60 and a Perkin Elmer 681 spectrophotometers (range 4000–200 cm^{-1}). ^1H NMR spectra were recorded on a Varian XR300 and a Bruker 200 spectrometers; chemical shifts are given in ppm with TMS as internal standard. Optical rotation values were measured with a Perkin Elmer 241MC polarimeter. The precursor $MoO_2(acac)_2$ was obtained from Fluka. All solvents were dried and distilled before use according to standard procedures. Ligand silane precursor $NH_2(CH_2)_3Si(OEt)_3$ was distilled before use. The inorganic supports were : Merck 60 (particle size 63–200 μm) and ultrastable Y zeolite (USY) prepared by steam calcination at 1300 K of partially ammonium exchanged NaY zeolite (SK–40 Union Carbide) followed by treatment with a 1N citric acid solution at 333 K for 30 minutes for "cleaning" material of extraframework species. After this, the zeolite was washed and dried. The main characteristics of the resultant zeolite are : unit cell size : 24–40 Å, bulk SiO_2/Al_2O_3 : 4.2, crystallinity : 95%.

Catalytic activity tests of the synthesized MoO_2(acac)L (13–17) for the selective epoxidation of 1–methylcyclohexene, styrene and 1–methylstyrene, as models of different substituted olefins, are shown.

Preparation of the Ligands

(S)–N–tert–Butylaminocarbonyl–2–pyrrolidinylmethanol (1a). To a solution of (S)–prolinol (2 g, 20 mmol) in methylene chloride (150 ml), cooled in an ice–bath, we added dropwise with efficient stirring a solution of tert–butylisocyanate (1.98 g, 20 mmol) under argon atmosphere. The reaction was monitored by t.l.c. using ethyl acetate–hexane (3:1) as eluent; after the reaction was completed ($\approx$ 1 h) the solvent was evaporated under reduced pressure to yield 1a (3.98 g). Yield : 100%; m.p. : 122–4 °C; $[\alpha]_D^{25}$ = –51.4° (acetone, 1). IR(cm^{-1}) : ν(NC=ON) : 1620(s), 1550(s); ν(OH) : 3340(s), 3150(m). ^{1}H–NMR(Cl$_3$CD) : δ = 4.62, 4.38 (broad, 1H+1H, NH, OH); 4.03 (m, 1H, CH); 3.6–3.3 (m, 2H, NCH$_2$); 3.24 (t, 2H, CH$_2$OH); 2.0–1.8 (m, 3H, H–3, H–4, H–4'); 1.55 (m, 1H, H–3'); 1.36 (s, 9H, CCH$_3$).

(S)–N–(3–Triethoxysilyl)propylaminocarbonyl–2–pyrrolidinylmethanol (1b). Following the procedure for 1a, 1b (6.94 g) was obtained starting from (S)–prolinol (2 g, 20 mmol) and (3–triethoxysilyl)propylisocyanate (4.94 g, 20 mmol) . Yield : 100%; $[\alpha]_D^{25}$ = –35.5° (methanol, 1). IR(cm^{-1}) : ν(NC=ON) : 1610(s), 1535(s); ν(OH) : 3300(s). ^{1}H–NMR(Cl$_3$CD) : δ = 5.0 (broad, 1H, OH); 4.08 (m, 1H, CH); 3.80 (q, 6H, CH$_2$OSi); 3.7–3.4 (m, 2H, H–5, H–5'); 3.29 (t, 2H, CH$_2$OH); 3.20 (q, 2H, NCH$_2$); 2.1–1.8 (m, 3H, H–3, H–4, H–4'); 1.7–1.5 (m, 3H, H–3', CH$_2$–CH$_2$Si); 1.24 (t, 9H, CH$_3$); 0.64 (t, 2H, CH$_2$Si).

(S)–N–tert–Butylaminocarbonyl–2–(1,1–diphenylmethyl)pyrrolidinylmethanol (2a). Following the procedure for 1a, 2a (280 g) was obtained starting from (S)–2–(1,1–diphenylmethyl)–pyrrolidinylmethanol [1] (200 mg, 0.8 mmol) and tert–butylisocyanate (80 mg, 0.8 mmol). Yield: 100%; m.p.: 163–4 °C; $[\alpha]_D^{25}$ = –105.5° (methanol, 1). IR(cm^{-1}) : ν(NC=ON) : 1630(s), 1520(s); ν(OH,NH) : 3460(s), 3280(s). ^{1}H–NMR(Cl$_3$CD) : δ = 7.5–7.2 (m, 10H, arom); 5.18 (dd, 1H, CH); 4.3 (broad, 1H NH); 3.12 (dd, 1H, H–5); 2.80 (m, 1H, H–5'); 2.2–1.9 (m, 3H, H–3, H–4, H–4'); 1.55–1.45 (m, 1H, H–3'); 1.30 (s, 9H, CCH$_3$).

(S)–N–(3–Triethoxysilyl)propylaminocarbonyl–2–(1,1–diphenylmethyl)pyrrolidinylmethanol (2b). Following the procedure for 1a, 2b (397 mg) was obtained starting from (S)–2–(1,1–diphenylmethyl)pyrrolidinylmethanol [8] (200 mg, 0.8 mmol) and (3–triethoxysilyl)propylisocyanate (197 mg, 0.8 mmol). Yield : 100%; $[\alpha]_D^{25}$ –104.5° (methanol, 1.03). IR(cm^{-1}) : ν(NC=ON) : 1615(s), 1545(s); ν(OH,NH) : 3340(s). ^{1}H–NMR(Cl$_3$CD) : δ = 7.5–7.2 (m, 10H, arom); 5.10 (dd, 1H, CH); 4.72 (broad, 1H, NH); 3.85 (q, 6H, CH$_2$OSi); 3.31 (m, 2H, NCH$_2$); 3.14 (dd, 1H, H–5); 2.82 (m, 1H, H–5'); 2.2–1.9 (m, 2H, H–3, H–4); 1.7–1.4 (m, 4H, H–3', H–4', CH$_2$CH$_2$Si); 1.22 (s, 9H, CH$_3$); 0.65 (m, 2H, CH$_2$Si).

(2S,4R)–N–tert–butylaminocarbonyl–trans–4–hydroxy–2–(1,1–diphenylmethyl)–pyrrolidinylmethanol (3a).

(2S,4R)–N–benzyl–trans–4–hydroxy–2–(1,1diphenylmethyl)pyrrolidinylmethanol
(6). Magnesium (925 mg, 38.1 mmol) and THF (75 ml) were poured in a flamed flask under ultrasound irradiation in argon atmosphere; a solution of bromobenzene (5.6 g, 38.1 mmol) in THF (10 ml) was dropwise added; the irradiation was maintained for 2 hours. After the magnesium was reacted a solution of (2S,4R)–N–benzyl–trans–4–hydroxy–2–methoxy–carbonylpyrrolidine (3.0 g, 12.7 mmol) in THF (5 ml) was added dropwise and the reaction mixture stirred for 18 hours. The reaction mixture was cooled, quenched with a saturated solution of ammonium chloride (if magnesium hydroxide precipitated, then solid NH_4Cl was added) and extracted with ethyl ether (150 ml). The organic extracts were successively washed with water (2x20 ml) and brine (25 ml) and the solvent evaporated under reduced pressure to yield **6**, that was used in next reaction without any purification. A sample was recrystallized from methanol–water (3:1). Yield : 78%; m.p. : 158–9 °C; $[\alpha]_D^{25} = +58.9°$ (methanol, 1). IR(cm^{-1}) : ν(CC–cycl) : 1495(s), 1450(s); ν(OH) : 3450(s). ^{1}H–NMR(Cl$_3$CD) : δ = 7.7–7.0 (m, 15H, arom.); 5.0 (broad, 1H, OH); 4.46 (t, 1H, CHOH); 4.31 (m, 1H, CHN); 3.42 (AB, 2H, CH$_2$Ph); 3.10 (dd, 1H, H–5); 2,52 (dd, 1H, H–5'); 1.9–1.7 (m, 2H, H–3, H–3').

(2S,4R)–trans–4–hydroxy–2–(1,1–diphenylmethyl)pyrrolidinylmethanol **(7)**. A mixture of (2S,4R)–N–benzyl–trans–4–hydroxy–2–(1,1–diphenylmethyl)–pyrrolidinyl–methanol (3.63 g, 10.12 mmol) in ethanol (20 ml) and 5% Pd/C (600 mg) was hydrogenated at 5 atm until total removing of benzyl group (t.l.c.). The reaction mixture was filtered on celite and the solvent eliminated under reduced pressure. The residue was dissolved in ethyl ether (100 ml) and acidified with hydrogen chloride (g) until pH=1 and the (2S,4R)–N–benzyl–trans–4–hydroxy–2–(1,1–diphenylmethyl)pyrrolidinylmethanol·HCl precipitated. The chlorhydrate was recrystallized from MeOH–ethyl ether and washed with ether. The purified salt was suspended in ethyl ether, basified with 2N sodium hydroxide under vigorous stirring. The aqueous layer was extracted with ethyl ether and the ethereal extracts were washed successively with water (2x20 ml) and brine (20 ml). The solution was concentrated to dryness and the residue was recrystallized from MeOH–water (3:1). Yield : 97%; m.p. : 123–4 °C; $[\alpha]_D^{25} = -114.9°$ (methanol, 1). IR(cm^{-1}) : ν(CC–cycl.) : 1495(s), 1450(s); ν(NH, OH) : 3400–3200 (s). ^{1}H–NMR(Cl$_3$CD) : δ = 7.9–7.3 (m, 10H, arom.); 4.69 (m, 1H, CHOH), 3.48 (dd, 1H, CHN); 3.52 (dd, 1H, H–5); 3.29 (dd, 1H, H–5'); 2.4–1.7 (m, 2H, H–3, H–3').

(2S,4R)–N–tert–butylaminocarbonyl–trans–4–hydroxy–2–(1,1–diphenyl–methyl)–pyrrolidinylmethanol **(3a)**. Following the procedure for **1a**, **3a** (274 mg) was obtained starting from (2S,4R)–trans–4–hydroxy–2–(1,1–diphenylmethyl)–pyrrolidinylmethanol (200 mg, 0.74 mmol) and tert–butylisocyanate (74 mg, 0.74 mmol). Yield : 100%; m.p. : 147–50 °C; $[\alpha]_D^{25} = -59.8°$ (acetone, 1). IR(cm^{-1}) : ν(NC=ON) : 1630(s), 1585(s); ν (OH,NH) : 3340 (s). ^{1}H–NMR(Cl$_3$CD) : δ = 7.5–7.2 (m, 10H, arom); 6.15, 4.40 (s+s, 1H+1H, NH, OH); 5.18 (m, 1H, CHOH); 4.40 (m, 1H, CHN); 3.31 (dd, 1H, H–5); 2.88 (dd, 1H, H–5'); 2.3–2.1 (m, 2H, H–3, H–3'); 1.30 (s, 9H, CCH$_3$).

(2S,4R)–N–(3–triethoxysilyl)propylaminocarbonyl–trans–4–hydroxy–2–(1,1–di–phenylmethyl)pyrrolidinylmethanol (3b). Following the procedure for **1a**, **3b** (382 mg) was obtained starting from (2S,4R)–trans–4–hydroxy–2–(1,1–diphenylmethyl)–pyrroli–

dinylmethanol (200 mg, 0.74 mmol) and (3–triethoxysilyl)propylisocyanate (182 mg, 0.74 mmol). Yield : 100%; $[\alpha]_D^{25}$ = –47.7° (methanol, 0.6). IR(cm^{-1}) : ν(NC=ON) : 1595 (s), 1540 (s); ν(OH,NH) : 3340(s). ^{1}H–NMR(Cl$_3$CD) : δ = 7.5–7.2 (m, 10H, arom); 6.12 (s, 1H, NH); 5.23 (m, 1H, CHOH); 4.40 (m, 1H, CHN); 3.85 (q, 6H, CH$_2$OSi); 3.32 (dd, 1H, H–5); 3.20 (q, 2H, NCH$_2$); 2.79 (dd, 1H, H–5'); 2.3–2.1 (m, 2H, H–3, H–3'.); 1.8–1.6 (m, 2H, CH$_2$–CH$_2$Si); 1.22 (t, 9H, CH$_3$); 0.65 (m, 2H, CH$_2$Si).

(2S,4R)–N–tert–butylaminocarbonyl–4–hydroxy–4–phenyl–2–(1,1–diphenylmethyl)pyrrolidinylmethanol (4,5a).

(S)–N–benzyl–4–oxo–2–methoxycarbonylpyrrolidine (**8**). A solution of dimethyl sulphoxide (1.7 g, 22 mmol) in methylene chloride (5 ml) was added to a cooled solution (–60 °C) of methylene chloride (25 ml) and oxalyl chloride (1.4 g, 11 mmol) in 2 minutes, followed by a 2 ml–solution of (2S,4R)–N–benzyl–*trans*–4–hydroxy–2–methoxycarbonyl-pyrrolidine (2.35 g, 9.43 mmol). After stirring for 1 hour at –60 °C, triethylamine (7 ml, 50 mmol) was added and then allowed to warm to room temperature. Water (50 ml) was added and the aqueous layer was extracted with an additional amount (50 ml) of methylene chloride. The organic extract was washed successively with water (20 ml) and brine (20 ml), dried over MgSO$_4$ and concentrated to dryness. The crude ketone was distilled to yield **8** (b.p.: 130 °C/0.3 torr). Yield : 86%; m.p. : 46–7 °C. $[\alpha]_D^{25}$ = –48.1° (methanol, 1). IR(cm^{-1}) : ν(COOMe : 1765(s); ν(C=O) : 1750 (s). ^{1}H–NMR(Cl$_3$CD) : δ = 7.5–7.2 (m, 5H, arom); 3.85 (m, 1H,) CHN); 3.92 (d, 1H, H–5); 3.75 (d, 1H, H–5'); 3.35, 3.02 (AB, 2H, CH$_2$Ph); 2.9–2.4 (m, 2H, H–3, H–3').

(2S,4R and 4S)–N–benzyl–4–hydroxy–4–phenyl–2–(1,1–diphenylmethyl)pyrrolidi-nylmethanol (**9, 10**). Magnesium (925 mg, 38.1 mmol) and THF (75 ml) were poured in a flamed flask under ultrasound irradiation in argon atmosphere; a solution of bromobenzene (5.6 g, 38.1 mmol) in THF (10 ml) was dropwise added; the irradiation was maintained for 2 hours. After the magnesium was reacted a solution of (2S)–1–benzyl–*trans*–4–oxo–2–methoxycarbonylpyrrolidine (**8**) (3.0 g, 12.7 mmol) in THF (5 ml) was added dropwise and the reaction mixture heated under reflux for 16 hours. The reaction mixture was cooled, quenched with a saturated solution of ammonium chloride (if magnesium hydroxide precipitated, then solid NH$_4$Cl was added) and extracted with ethyl ether (150 ml). The organic extract was successively washed with water (2x20 ml) and brine (25 ml) and the solvent evaporated under reduced pressure to yield a mixture of two diasteromers (70%) in a ratio **9:10** (30/70). The diastereomeric mixture was chromatographed on silica gel using ethyl acetate:hexane (10:1) as eluent to yield (2S,4R)–**9** (Rf=0.35; 600 mg) and (2S,4S)–**10** (Rf= 0.22; 2.10 g), that were used in the next reactions without any purification. Samples were recrystallized from methanol–water (3:1).

9: M.p. : 220–2 °C; $[\alpha]_D^{25}$ = +94.4° (methanol, 0.5). IR(cm^{-1}) : ν(CC–cycl.) : 1500 (s), 1450 (s); ν(OH) : 3460 (s). ^{1}H–NMR(Cl$_3$CD) : δ = 7.9–6.9 (m, 20H, arom.); 4.35 (dd, 1H, CHN); 3.70 (s, 1H, OH); 3.18 (dd. 1H, H–5); 3.4–3.0 (AB, 2H, CH$_2$Ph); 2.83 (dd, 1H, H–5'); 2,75 (dd, 1H, H–3); 2.15(m, 1H, H–3').

10: M.p. : 200–3 °C; $[\alpha]_D^{25}$ = +31.5° (methanol, 1). IR(cm^{-1}) : ν(CC–cycl.) : 1495 (s), 1450 (s); ν(OH) : 3440 (s). ^{1}H–NMR(Cl$_3$CD) : δ = 7.9–6.9 (m, 20H, arom.); 5.2 (s, 1H, OH); 4.73 (dd, 1H, CHN); 3.7–3.4 (AB, 2H, NCH$_2$Ph); 3.28 (dd, 1H, H–5); 2,95 (dd, 1H, H–5'); 2.3–2.0 (m, 2H, H–3, H–3').

(2S,4R)–hydroxy–4–phenyl–2–(1,1–diphenylmethyl)pyrrolidinylmethanol **(11)**. Following the procedure for **7**, **11** was obtained starting from *(2S,4R)–N*–benzyl–4–hydroxy–4–phenyl–2–(1,1–diphenylmethyl)pyrrolidinylmethanol **(9)** (200 mg, 0.46 mmol) and 20 mg Pd/C. Yield : 100%; m.p. : 198–204 °C; $[\alpha]_D^{25}$ = –42.3° (methanol, 1). IR(cm^{-1}) : ν(CC–cycl.) : 1495 (s), 1450 (s); ν(OH) : 3400 (s). ^{1}H–NMR(DMSO–d_6–500 MHz) : δ = 7.7–7.1 (m, 15H, arom); 5.00 (broad, 1H, NH); 4.32 (m, 1H, CHN); 2.95 (m, 1H, H–5); 2.86 (dd, 1H, H–5'); 2.22 (m, 1H, H–3); 1.50 (dd, 1H, H–3').

(2S,4S)–4–hydroxy–4–phenyl–2–(1,1–diphenylmethyl)pyrrolidinylmethanol   **(12)**. Following the procedure for **7**, **12** was obtained starting from *(2S,4S)–N*–benzyl–4–hydroxy–4–phenyl–2–(1,1–diphenylmethyl)pyrrolidinylmethanol **(10)** (3.1 g, 7 mmol) and 600 mg Pd/C. Yield : 100%; M.p. : 160–1 °C; $[\alpha]_D^{25}$ = –16.6° (methanol, 1). IR(cm^{-1}) : ν (CC–cycl.) : 1495 (s); 1450 (s); ν(OH,NH) : 3390 (s). ^{1}H–NMR(DMSO–d_6–500 MHz) : δ = 7.6–7.0 (m, 15H, arom); 5.03 (t, 1H, NH); 4.05 (m, 1H, CHN); 3.05 (m, 1H, H–5); 2.75 (m, 1H, H–5'); 2.21 (m, 1H, H–3); 1.79 (m, 1H, H–3').

(2S,4R)–N–tert–butylaminocarbonyl–4–hydroxy–4–phenyl–2–(1,1–diphenylmethyl)pyrrolidinylmethanol **(4a)**. Following the procedure for **1a**, **4a** (260 mg) was obtained starting from *(2S,4S)*–4–hydroxy–4–phenyl–2–(1,1–diphenylmethyl)–pyrroli–dinyl–methanol **(11)** (200g, 0.6 mmol) and *tert*–butylisocyanate (60 mg, 0.6 mmol). Yield : 100%; $[\alpha]_D^{25}$ = +84.6° (methanol, 0.5). IR(cm^{-1}) : ν(NC=ON) : 1630 (s), 1540 (s); ν(CC–cycl.) : 1495 (m); 1450 (s); ν(OH,NH) : 3400 (s). ^{1}H–NMR(Cl$_3$CD) : δ = 7.8–7.4 (m, 15H, arom); 5.10 (s, 1H, NH); 5.20 (m, 1H, CHN); 3.86 (d, 1H, H–5); 3.28 (d, 1H, H–5'); 2.4–2.2 (m, 2H, H–3, H–3'); 1.32 (s, 9H, CCH$_3$).

(2S,4S)–N–tert–butylaminocarbonyl–4–hydroxy–4–phenyl–2–(1,1–diphenylmethyl)pyrrolidinylmethanol **(5a)**. Following the procedure for **1a**, the compound 5a (260 mg) was obtained after 1 hour, starting from *(2S,4S)*–4–hydroxy–4–phenyl–2–(1,1–diphenylmethyl)pyrrolidinylmethanol **(12)** (200 mg. 0.6 mmol) and *tert*–butylisocyanate (60 mg, 0.6 mmol). Yield : 100%; m.p. : 180–3 °C; $[\alpha]_D^{25}$ = +56.6° (methanol, 1). IR(cm^{-1}) : ν (NC=ON) : 1630 (s), 1530 (s); ν(CC–cycl.) : 1495 (m); 1450 (s); ν(OH,NH) : 3400 (s). ^{1}H–NMR(DMSO–d_6) : δ = 7.8–7.4 (m, 15H, arom); 5.80 (s, 1H, NH); 5.33 (m, 1H, CHN); 4.10 (d, 1H, H–5); 3.21 (d, 1H, H–5'); 2.5–2.3 (m, 2H, H–3, H–3'); 1.38 (s, 9H, CCH$_3$).

(2S,4R)–N–3–(triethoxysilyl)propylaminocarbonyl–4–hydroxy–4–phenyl–2–(1,1–diphenylmethyl)pyrrolidinylmethanol (4b). Following the procedure for **1a**, the compound **4b** (348 mg) was obtained after 6 hours, starting from *(2S,4R)*–4–hydroxy–4–phenyl–2–(1,1–diphenylmethyl)pyrrolidinylmethanol **(11)** (200g, 0.6 mmol) and (3–triethoxysilyl)propylisocyanate (148 mg, 0.6 mmol). Yield : 100%; $[\alpha]_D^{25}$ = +62.8° (methanol, 1). IR(cm^{-1}) : ν(NC=ON) : 1610 (s), 1540 (s); ν(OH,NH) : 3365 (s). ^{1}H–NMR(C$_6$D$_6$) : δ = 7.8–7.0 (m, 15H, arom); 5.35 (m, 1H, CHN); 3.85 (q, 6H, CH$_2$OSi); 3.63 (d, 1H, H5); 3.22 (m, 2H, CH$_2$N); 3.02 (dd, 1H, H–5'); 2.4–2.1 (m, 2H, H–3, H–3'); 1.20 (s, 9H, CH$_3$); 0.64 (m, 2H, CH$_2$Si).

(2S,4R)–N–3–(triethoxysilyl)propylaminocarbonyl–4–hydroxy–4–phenyl–2–(1,1–diphenylmethyl)pyrrolidinylmethanol (5b). Following the procedure for **1a**, the compound **5b** (690 mg) was obtained after 6 hours, starting from *(2S,4R)*–4–hydroxy–4–phenyl–2–(1,1–diphenylmethyl)pyrrolidinylmethanol **(12)** (400 g, 1.2 mmol) and 3–tri–

ethoxysilylpropylisocyanate (300 mg, 1.2 mmol). Yield : 100%; $[\alpha]_D^{25}$ = +42.5° (methanol, 0.55). IR(cm^{-1}) : ν(NC=ON) : 1610 (s), 1535 (s); ν(OH,NH) : 3380 (s). ^{1}H–NMR(C$_6$D$_6$) : δ = 7.8–7.0 (m, 15H, arom); 5.40 (m, 1H, CHN); 3.85 (q, 6H, CH$_2$OSi); 3.77 (d, 1H, H5); 3.20 (m, 2H, CH$_2$N); 2.70 (m, 1H, H–5'); 2.45 (m, 1H, H–3); 2.35 (m, 1H, H–3'); 1.18 (s, 9H, CH$_3$); 0.64 (m, 2H, CH$_2$Si).

Preparation of Dioxomolybdenum Complexes.

General Procedure. A solution of the corresponding ligand **1–5** (0.43 mmol) in methylene chloride or cyclohexane (10–15 ml) was dropwise added to a stirred solution of MoO$_2$(acac)$_2$ (139 mg, 0.43 mmol) in the same solvent (50 ml) in a Schlenk type flask under argon atmosphere. A white precipitate was formed, the reaction mixture was stirred for 1–5 hours at r.t. (CH$_2$Cl$_2$) or 70 °C (cyclohexane). The crude solid complex was filtered, washed with ethyl ether or hexane. Significant analytical data and yields are shown in table 1.

Table 1.

Ligand (L)	Complex*	IR (cm^{-1})	1H-NMR (δ, ppm)	Yield (%)
1a	**13a**	ν(CO) 1630,1570 ν(MoO) 940,910	5.1(dd, 1H, CHN); 5,7, 2.0 (s+s, 1H+6H, acac); 1.2(s, 9H, CCH$_3$)	72
2a	**14a**	ν(CO) 1600,1520 ν(MoO) 940,910	5.0(dd, 1H, CHN); 5,3, 2.0(s+s, 1H+6H, acac); 1.2(s, 9H, CCH$_3$)	90
3a	**15a**	ν(CO) 1710,1600 ν(MoO) 930,900	5.2(dd, 1H, CHN); 5,7, 2.2(s+s, 1H+6H, acac); 1.2(s, 9H, CCH$_3$)	86
4a	**16a**	ν(CO) 1730,1600 ν(MoO) 940,910	5.1(dd, 1H, CHN); 5,7, 2.0 (s+s, 1H+6H, acac); 1.2(s, 9H, CCH$_3$)	60
5a	**17a**	ν(CO) 1720,1600 ν(MoO) 940,910	5.1(dd, 1H, CHN); 5,7, 2.1 (s+s, 1H+6H, acac); 1.2(s, 9H, CCH$_3$)	84
1b	**13b**	ν(CO) 1630,1550 ν(MoO) 940,910	5.1(dd, 1H, CHN); 5,65, 2.0 (s+s, 1H+6H, acac); 1.2(s, 9H, CCH$_3$)	60
2b	**14b**	ν(CO) 1595,1520 ν(MoO) 940,910	5.1(dd, 1H, CHN); 5,4, 2.0 (s+s, 1H+6H, acac); 1.2(s, 9H, CCH$_3$)	88
3b	**15b**	ν(CO) 1710,1600 ν(MoO) 940,910	5.1(dd, 1H, CHN); 5,7, 2.1 (s+s, 1H+6H, acac); 1.2(s, 9H, CCH$_3$)	82
4b	**16b**	ν(CO) 1730,1605 ν(MoO) 940,910	5.1(dd, 1H, CHN); 5,7, 2.0 (s+s, 1H+6H, acac); 1.2(s, 9H, CCH$_3$)	54
5b	**17b**	ν(CO) 1720,1600 ν(MoO) 940,910	5.1(dd, 1H, CHN); 5,7, 2.1 (s+s, 1H+6H, acac); 1.2(s, 9H, CCH$_3$)	81

*All Mo-complexes have shown C,H,N and Mo analyses concordant with their corresponding theoretical values.

Anchoring of Transition Metal Complexes.

General Procedure [6]. A MoO$_2$(acac)L–complex described previously **(13b–17b)** (0.15 mmol) in dry CH$_2$Cl$_2$ (10 ml) was added to a suspension of the elected inorganic support (1 g), previously dried at 160 °C/0.01 torr for 4 hours, in toluene (50 ml) and the mixture was stirred for 24 hours at room temperature in argon atmosphere. The solid was filtered off, extracted in a Soxhlet apparatus with CH$_2$Cl$_2$/ ethyl ether (1:1) for 24 hours, in order to remove non–supported complex, and dried at room temperature under reduced pressure. These catalysts are solids stable in air and moisture for at least three months without significant changes and contain 1–2% of metal (atomic absorption).

Catalytic Experiments.

The oxidation reactions of alkenes were carried out under argon atmosphere in a 50 ml flask containing CH$_2$Cl$_2$ (25 ml), catalyst (0.05 mmol of MoO$_2$(acac)L–complex), alkene (10.4 mmol) and *tert*–butyl hydroperoxide (dry solution in CH$_2$Cl$_2$ [9]) at room or solvent reflux temperature. The products formed were analyzed by g.l.c. by comparison of their retention time with authentic samples. The *ee*^{s} were measured by g.l.c. on a (10%)permethylcyclodextrin in OV1701 silicone capillary 25m–column [10] and by NMR using (hfc)$_3$Eu, as resolving reagent, and mathematical numerical integration of the corresponding signals, when they were not totally resolved

RESULTS AND DISCUSSION

In this paper we have prepared a series of MoO$_2$(acac)L **(13–17a,b)**, where **L** is a bidentate N,O– or O,O–ligand derived from (L)–proline and (L)–*trans*–4–hydroxyproline with an growing number of bulky substituents in the environment of catalytic center. The synthesis of ligands **1–5a,b** were carried out by classical organic chemistry procedures as shown in schemes 2 and 3, and the structures of compounds was assured by spectroscopic and analytical methods. The introduction of N–carbamoyl substituent containing an triethoxysilyl group, adequate for anchoring on support, was performed clean and simply by treatment with the corresponding alkylisocyanate at r. t. in quantitative yields without any hydrolysis of Si–OEt bonds.

All complexes were prepared via ligand exchange starting with MoO$_2$(acac)$_2$ and isolated as solids. The IR spectra show the presence of two bands at 940–910 cm^{-1} which correspond to two *cis* Mo=O stretching vibrations as well as bands of acetylacetonate group and the corresponding chiral ligand. The elemental analysis and ^{1}H NMR spectra confirm the substitution of only one molecule of acac by a pyrrolidine ligand. The NMR spectra show signals for acetylacetonate and pyrrolidine ligands, which are essentially at the same chemical shift that the initial Mo–complex and the respective free ligands presented. The stability to moisture and air were increased when the steric constrains of the ligands increased but these free Mo–complexes tend to dimerize for giving the non–useful Mo(V)– species which are not reactive in oxidation reactions. This tendency was eliminated by anchoring into USY zeolites, and the supported complexes may be safety manipulated in the air without inactivation.

Catalytic Oxidation of Olefins.

The free and supported Mo–complexes were tested in epoxidation of 1–methylcyclohexene, styrene and α–methyl styrene using TBHP as oxidation reagent under anhydrous conditions. In table 2 the comparative results are shown for the 1–methyl-cyclohexene and the turnover numbers (TOR) calculated in $mole_{substrate}/ mole_{catalyst} \cdot h^{-1}$ for the linear maximum rate (essentially 0–90% conversion).

Table 2.

Catalyst	Conv. (%)	Epoxide (%)	Diol (%)	Ketone (%)	TOR $(mol_s/mol_c \cdot h^{-1})$	ee (%)
13a	92	85	<5	<1	1.48	<5
Sil-13b	90	27	54	18.8	2.80	-
Zeol-13b	93	83	10	7.0	80.5	6
14a	98	70.8	23.3	5.9	2.73	6
Sil-14b	98	80	15	5.0	5.20	6
Zeol-14b	95	95	<5	<1	84.5	10
15a	90	92	8	<1	10.4	5
Zeol-15b	90	89	8.5	3.0	28.2	5
16a	90	92	8	<1	10.4	5
Zeol-16b	90	97.8	1.2	1.0	127	36
17a	95	91.6	6.0	2.4	45	4
Zeol-17b	97	98.2	1.0	1.8	149	24

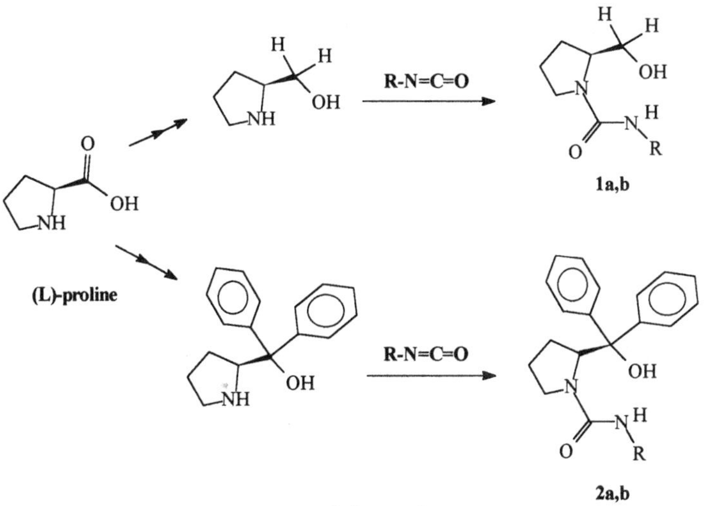

Scheme 2.

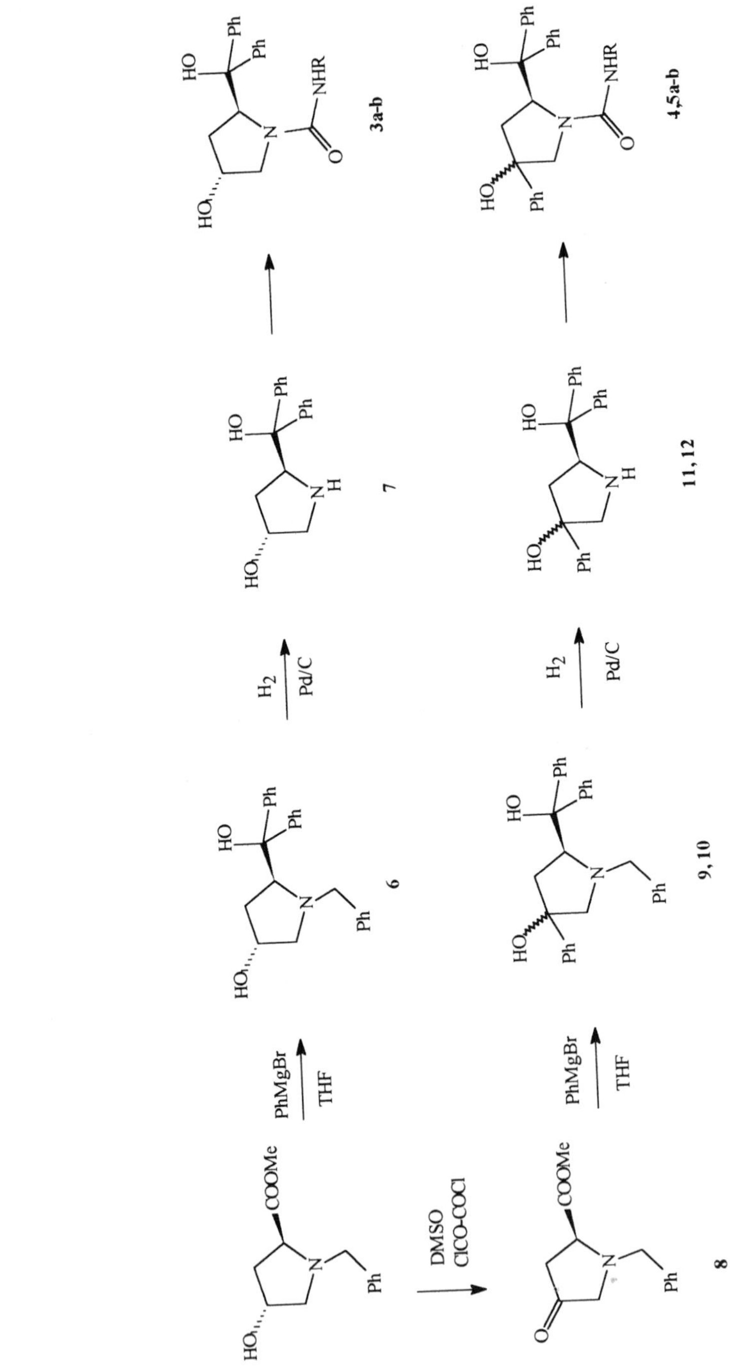

Scheme 3.

a: R = t-Bu; b: R = (CH₂)₃Si(OEt)₃

When the Mo–complexes were anchored into a matrix, the TOR numbers for epoxidation of 1–methylcyclohexene at 40 °C increased, more remarkable in the zeolite supported catalysts where the rate was enhanced up to 50 times with respect to the homogeneous ones, probably due to concentration and electric interactions with the surface and an increase in the stability of metallic center in the reaction conditions. The same oxidation, at room temperature, looks similarly, but the selectivity to the epoxide was lower (68–92%). The increase of enantioselectivity when Mo–complexes are incorporated into zeolites, is a systematic finding, but with the used support and conditions, only ee = 6–36 % were obtained. The epoxidations of styrene and α–methyl styrene follow a same pattern, but unfortunately the epoxides were overoxidated to benzaldehyde, benzoic acid or acetophenone respectively by C–C bond breaking and the selectivity were 70–80%. Studies on the mechanism of the reaction for free and supported catalysts are in progress.

ACKNOWLEDGMENTS.

Financial support by the Dirección General de Investigación Científica y Técnica (Project MAT91–1097–C02–02) is gratefully acknowledged.

REFERENCES

[1]   V. Schurig, B. Koppenhoefer, W. Bürkle, J. Org. Chem, 45 (1980) 538.
[2]   J.W. Scott, in "Asymmetric Synthesis", J.D. Morison, Ed., Academic Press, New York (1984) vol. 4, pp. 1–226.
[3]   a) K.B. Sharpless, S.S. Woodard, M.G. Finn, Pure & Appl. Chem., 55 (1983) 1823.
      b) T. Katsuki, K.B. Sharpless, J. Am. Chem. Soc.,102 (1980) 5974.
[4]   K.A. Jorgensen, Chem. Rev., 89 (1989) 43.
[5]   S. Coleman–Kammula, E.T. Duim–Ksalstra, J. Organomet. Chem., 246 (1983) 53.
[6]   A. Corma, M. Iglesias, C. del Pino, F. Sanchez, J. Organomet. Chem., 431 (1992) 233.
[7]   F.R. Hartley, in "Supported Metal Complexes", D. Riedel, Netherlands, 1985.
[8]   D. Enders, H. Kipphardt, P. Gerdes, L.J. Brena–Valle, V. Bhushan, Bull. Soc. Chim. Belg., 97 (1989) 691.
[9]   K.B. Sharpless, R. Verhoeven, Aldrichimica Acta, 12 (1979) 63.
[10] Chiral column was prepared in our Chromatography Dpt. (Drs. J. Sanz and M.I. Martinez).

EFFECT OF SULFOXIDE LOADING ON THE SELECTIVITY AND ACTIVITY OF ZEOLITE Y FOR DEHYDRATION REACTIONS

D.Bethell[1], S.Feast[2], G.J.Hutchings[2], F. King[3], P.C.B.Page[1] and S.P.Saberi[2].

[1]Chemistry Department
[2]Leverhulme Centre For Innovative Catalysis
Liverpool University
P.O. BOX. 147,
Liverpool L69 3BX, United Kingdom
[3]ICI Katalco
P.O. BOX 1, Billingham
Cleveland T823 1LB, United Kingdom

SUMMARY

The incorporation of chiral sulfoxides 1,3–dithiane oxide, 2–methyl–1,3–dithiane oxide and 2–phenyl–1,3–dithiane oxide– into zeolite Y by incipient wetness is being studied as a system for heterogeneous enantioselective catalysis. The system is shown to be stable under reaction conditions by solid and solution state NMR spectroscopy, electron ionisation mass spectrometry (EIMS) and X–ray diffraction (XRD). Modified zeolite Y has been catalytically tested for kinetic selectivity in the dehydration of butan–2–ol, giving enhanced reactivity and selectivity to but–2–ene.

INTRODUCTION

The synthesis of enantiomerically pure compounds has become increasingly important as our awareness of the biological implications of chirality grows [1]. Many drugs, pesticides and food additives are chiral and therefore, exist in two enantiomeric forms. The fundamental aspects and biological significance of chirality are starting to be understood. An poignant example of the effects of chirality being the drug thalidomide in which the left handed isomer is a powerful tranquilliser, whilst the right handed isomer is a powerful teratogen causing severe handicaps. Cases such as thalidomide have led to much research and in the USA [2] legislation on enantiomeric purity. The need to produce large quantities of enantiomerically pure compounds is a challenge for heterogeneous catalysis, an area in which many reviews have been published [3,4].

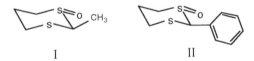

I   II

Figure 1. Sulfoxides used to enhance selectivy of zeolite Y. 2–methyl–1,3–dithiane oxide (I) and 2–phenyl–1,3–dithiane oxide (II).

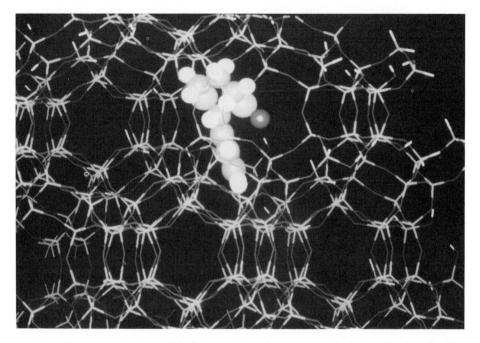

Figure 2. A frame of 2–phenyl–1,3–dithiane oxide sat in an energy minimised position in zeolite Y.

Our approach to the problem is to modify zeolites with chiral sulfoxides (figure 1), aiming to achieve enantioselectivity by combining the shape selectivity of the zeolite and the chirality of the sulfoxide. There are two main techniques for incorporating the modifiers into a zeolite structure; either as a template in the synthesis gel or post synthesis, often by sorption. This first route is exemplified in a patent, which describes the use of optically active amines as templates in the synthesis of a high silica zeolite [5]. However, no chiral character was reported for the zeolite. More recently Corma *et al.* [6] have demonstrated the efficacy of the second method by successfully anchoring a rhodium complex with *N– –* based chiral ligands onto modified USY zeolite, achieving enantioselective hydrogenation of prochiral amines.

Both approaches have also been tried with respect to modifying zeolite Y with chiral sulfoxides. However, we have found that although zeolite Y crystallises in the presence of dithiane, the resulting catalyst is not very active, whereas zeolite Y modified by incipient wetness demonstrates increased activity and selectivity for catalytic dehydration of butan–2–ol.

EXPERIMENTAL

Zeolite Y (2.0125 g) was treated with a solution of 2–methyl–1,3–dithiane oxide (0.1439 g) in de–ionised water (30 ml) for 2 hours at 50 °C. The modified zeolite was filtered, dried at 100 °C then pressed and sieved to give particles of 1000–600 μm for micro reactor testing. Control samples of pure zeolite Y were prepared similarly, omitting the sulfoxide from the solution. XRD data was collected on a Hilton–Brooks modified Philips 1050W diffractometer with a Cu Kα source (40 keV and 20 mA). All MAS NMR spectra were acquired on a Bruker 400 MHz spectrometer. ^{27}Al, ^{29}Si and ^{13}C frequencies were 104.22, 79.5 and 100.13 MHz respectively. Zirconia rotors were spun at 4 kHz for ^{29}Si and ^{13}C MAS, and 12 kHz for ^{27}Al MAS. The ^{29}Si and ^{13}C MAS experiments were acquired with a π/2 pulse and the ^{27}Al was acquired using a π/12 pulse. The ^{13}C solution state NMR spectrum was acquired on a Bruker AMX400 spectrometer, referenced to TMS.

BIOSYM molecular modelling was carried out on an Silicon Graphics Iris Indigo workstation, using Monte Carlo Docking and energy minimisation programmes to visualise the interaction of the sulfoxides within the zeolite framework.

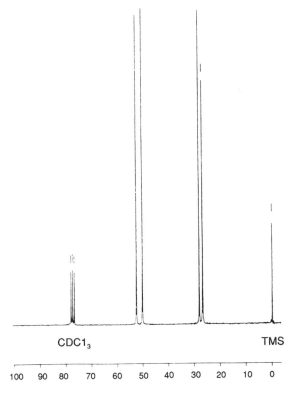

Figure 3. ^{13}C solution state nmr of dithiane oxide inCDCl₃, referenced to TMS.

RESULTS AND DISCUSSION

The adsorption of the sulfoxides inside the pores of zeolite Y has been studied using computer simulation techniques. Molecular dynamics procedures have been used to locate positions of minimum energy of the sulfoxides inside the zeolite. A Monte–Carlo docking programme [5] was then employed to investigate these sites. Figure 2 shows one of the possible docked positions of 2–phenyl–1,3–dithianeoxide in zeolite Y. The sulfoxide fits easily into the channel structure without blocking the pores. The phenyl ring is protruding into the supercage, with the oxygen from the sulfoxide also directed into the supercage.

^{13}C NMR has been employed to demonstrate that the sulfoxides remain intact inside the zeolite. The solution state nmr (figure 3) of dithiane oxide (figure 4) shows 4 peaks: C_2, C_6, C_4 and C_5 at 52, 50, 28 and 26 ppm respectively. The ^{13}C MAS NMR of dithiane loaded into zeolite Y by incipient wetness (figure 5) shows two main peaks at 51, 28 and 26 ppm corresponding to C_2, C_6 and C_4, C_5 of intact dithiane oxide.

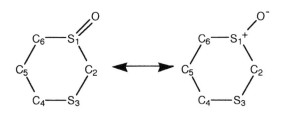

Figure 4. Labelled conformational isomers of 1,3–dithiane oxide.

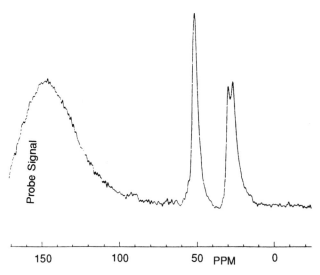

Figure 5. ^{13}C MAS nmr of dithiane loaded into zeolite Y by incipient wetness. The spectrum is referenced to TMS.

The presence of sulfoxide has been shown by XRD (figure 6) not to affect the crystallinity of zeolite Y.

Micro reactor studies of the sulfoxide modified zeolite Y have shown greatly increased activity; the modified catalyst is active at 50 °C lower temperature, and selectivity to trans–but–2–ene, from 40.6% to 53.8%.

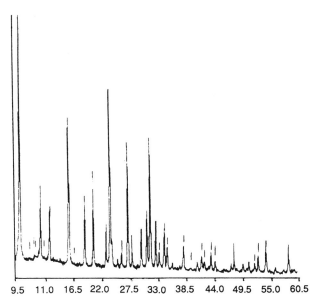

Figure 6. XRD pattern of zeolite Y with 10% loading of dithiane–1–oxide.

CONCLUSIONS

The characterisation experiments show that the sulfoxides are stable within the acidic framework of the zeolite. Molecular modelling studies have shown that the sulfoxides will not destroy the zeolites porous structure. Micro reactor tests suggest that this system promotes activity and selectivity in dehydration reactions. Current work involves using enantiomerically pure sulfoxides to impregnate the zeolite.

ACKNOWLEDGEMENTS

We would like to thank Dr. Rob Bell (Royal Institution) for help with the molecular modelling, Dr. Tom Garrett for the sulfoxides and Dr. Bodo Zibrowius for very useful discussions about MAS NMR. The authors gratefully acknowledge SERC and ICI for the funding of this work.

REFERENCES

[1]   E. Juaristi. *Introduction to Stereochemistry and Conformational Analysis,* Wiley, 1991.
[2]   S. C.Stinson. *Chem. Eng. News,* Sept.27, 1993.
[3]   H. Blaser and M.Muller, *Heterogeneous Catalysis and Fine Chemicals II,* (eds. M. Guisnet e.a.) Elsevier 1991, p 73.
[4]   Y. Izumi, *Advances in Catalysis,* 1983, **32,** 215.
[5]   C.F. Freeman, C.R.a.Catlow, J.M.Thomas, S.Brode. *Chem. Phys. Lett.,* 1991, **186**, 136.

NOVEL CHIRAL RUTHENIUM–BIS(OXAZOLINYL)PYRIDINE AND BIS-(OXAZOLINYL)–BIPYRIDINE COMPLEXES: ASYMMETRIC CYCLOPROPANATION OF STYRENE WITH DIAZOACETATES

Hisao Nishiyama,* Hideki Matsumoto, Soon–Bong Park, and Kenji Itoh

School of Materials Science
Toyohashi University of Technology
Tempaku–cho
Toyohashi 441, Japan

INTRODUCTION

Recent development for the asymmetric cyclopropanation has been dependent on new chiral copper catalysts having several C_2–symmetrical nitrogen ligands, such as *semicorrins* and *bis–oxazolines* [1]. We have developed a C_2–symmetrical ligand bis(oxazolinyl)–pyridine (*pybox*) and bis(oxazolinyl)–bipyridine (*bipymox*) for the rhodium catalyzed hydrosilylation of ketones [2,3]. Although pybox would be applicable to the asymmetric cyclopropanation, we could not find any significant results in the asymmetric induction with copper(I) and rhodium(II) complexes. However we discovered an efficient system with pybox and ruthenium(II) complex for the asymmetric cyclopropanation of olefins with diazoacetates.

Ruthenium catalyzed cyclopropanation with diazoacetates was first reported in 1980 by Hubert and Noels followed by Doyle indicating that the catalytic activity of ruthenium catalysts for the cyclopropanation is nearly comparable to or lower than the one of copper and rhodium catalysts [4]. To our knowledge, this is the first report of the asymmetric version of the ruthenium catalyzed cyclopropanation.

RESULTS AND DISCUSSION

[Ru(II)Cl$_2$(*p*–cymene)]$_2$ [5] **1** was adopted as readily available and stable precatalyst. To a dark purple solution prepared from **1** (1 mol%) and pybox–(*S,S*)–*ip* **2** (*ip* = *iso*–Pr) (4 eq to Ru) in dichloromethane under inert atmosphere, excess of styrene (15 mmol) was added. Then *tert*–butyl diazoacetate (3.0 mmol) in dichloromethane was slowly added for ca. 8 h at 20–25 °C. Chromatographic separation gave the *trans*–cyclopropane **3** and the *cis*–isomer **4** in 81% (**3:4** = 97:3) with high enantioselectivities, 94% ee (*1R,2R*) and 85% ee (*1R,2S*) respectively.

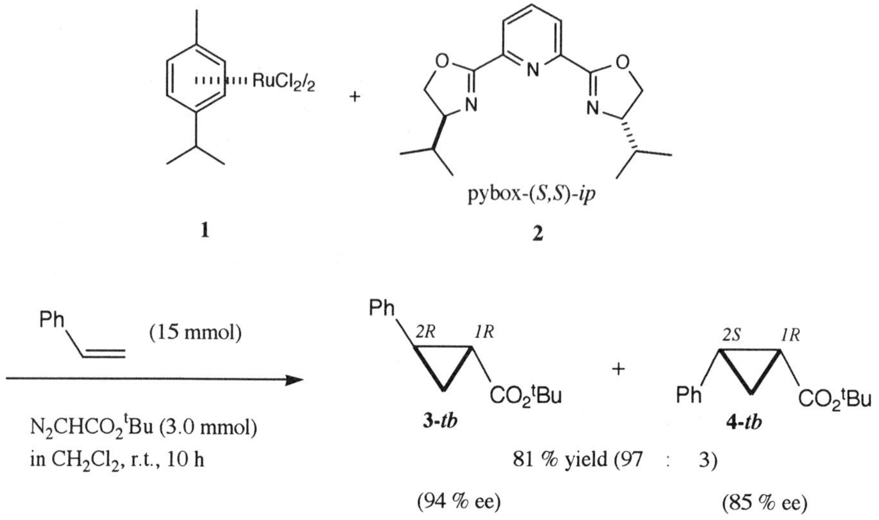

Scheme 1.

Results with other diazoacetates and pybox were summarized in table 1. Exclusive formation of the *trans*–isomers was observed with the increase of the bulkiness of the ester groups, accompanied with the increase of the enantioselectivity up to 95% ee. The higher turn over number 100 (**1**, 0.5 mol%) slightly decreased the enantioselectivity. Two equivalents of pybox ligand to ruthenium was sufficient for asymmetric induction. The effect of the substituent on pybox (*sb* = sec–Bu, *tb* = tert–Bu, *ph* = phenyl) in place of *iso*–propyl group was also examined under the same conditions. However it is proved that the *iso*–propyl group on pybox gave much better results comparing to those of *tert*–butyl or phenyl groups on pybox, in good contrast to the *tert*–butyl substituent of the reported bidentate ligands such as *semicorrin* and *bis(oxazoline)*.

We observed that the maleate rather than the fumarate was obtained as a major by–product (ca. 5 – 20% yields) during the cyclopropanation. Therefore we examined the dimerization of the diazoacetates in the absence of styrene under the same conditions as described above. The ratios of the *cis*– and *trans*–olefins increased with the bulkiness of the ester groups up to 92:8 for the *tert*–butyl diazoacetate (scheme 2)(table 2). Similar phenomena were recently reported for the reaction of ruthenium–tetramesitylporphyrin (*tmp*) and ethyl diazoacetate [6].

As a hypothetical intermediate and a transition state, a carbene complex of ruthenium(II)–pybox might be proposed (scheme 3). The prochiral *re*–face of the carbene center was selectively attacked by styrene to give the (*1R*)–configuration for both *trans*– and *cis*–phenylcyclopropane derivatives, **3** and **4**.

Treatment of the dark purple solution prepared from **1** and **2** with CO (1 atm) gave only an air–stable complex, *trans*–Cl$_2$(CO)Ru(pybox) **5** in 92% yield. The *trans*–configuration of **5** was confirmed by the symmetry of the ^{1}H NMR pattern. The stereochemistry of **5** is suggestive for the geometry of the intermediary carbene complex. The hypothetical active species, [RuCl$_2$–pybox], from **1** and **2** possesses one vacant coordination site *trans* to the pyridine nitrogen of pybox. A similar activity of one vacant site was reported for Rh$_2$(OCOR)$_4$ catalysts [7]. The carbonyl complex **5** itself has no catalytic activity for the cyclopropanation below 40 °C because of the saturation of metal coordination by CO. However **5** catalyzed the reaction of styrene and ethyl diazoacetate at 75 °C in 1,2–dichloroethane to give a mixture of **3** and **4** in 60% yield (69:31) but lower % ee's, 14% (*1R,2R*) and 4% (*1R,2S*) respectively. The corresponding monotriflate complex **6**, generated in–situ by treatment of **5** with AgTfO (TfO = trifluoromethanesulfonate), exhibited the catalytic activity at 40 °C to give **3** and **4** (55:45) in 38% yield, 44% ee (*1S,2S*) and 3% ee (*1R,2S*) respectively.

Table 1. Catalytic cyclopropanation of styrene with diazoacetates and Ru–pybox systems.[a]

pybox-	N$_2$CHCO$_2$R R =	product: yield (%)	ratio trans:cis	% ee trans	cis	abs config
(*S,S*)-ip	Me	68	90:10	86	75	*R*
	Et	69	92: 8	88	78	*R*
	i-Pr	65	92: 8	91	72	*R*
	t-Bu	81	97: 3	94	85	*R*
	d-Ment	85	95: 5	86	95	*R*
	l-Ment	87	95: 5	95	76	*R*
	Et[b]	66	92: 8	89	75	*R*
	Et[c]	40	88:12	83	61	*R*
(*S,S*)-sb	*t*-Bu	86	97: 3	91	73	*R*
(*S,S*)-tb	Et	11[d]	58:42	11	9[e]	*R*
(*R,R*)-ph	Et	20[d]	65:35	86	14	*S*

[a] [RuCl$_2$(*p*-cymene)]$_2$ (0.03 mmol), pybox (0.24 mmol), CH$_2$Cl$_2$ (2 mL), styrene (15 mmol), diazoacetate (3.0 mmol, ca. 1 *N* in CH$_2$Cl$_2$), 20~25 °C (addition for ca. 8 h then stirring for ca. 2 h. Isolated yields. The % ee's were determined by chiral capillary GLPC (Astec B-DA) of the corresponding methyl ester. The absolute configurations were determined by the signs of optical rotation: *R* (*trans-1R,2R*: *cis-1R, 2S*); *S* (*trans-1S, 2S*: *cis-1S,2R*).
[b] pybox (0.12 mmol, 2 eq to Ru).
[c] [RuCl$_2$(*p*-cymene)]$_2$ (0.015 mmol, *ton* 100), pybox (0.12 mmol).
[d] Ethyl diazoacetate was remained unreacted at 25 °C for 10 h: convertion < ca.50 %. [e] (*1S,2R*)

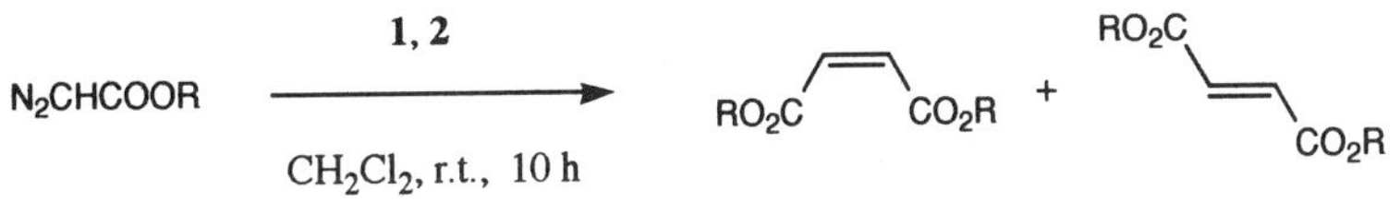

N$_2$CHCOOR $\xrightarrow[\text{CH}_2\text{Cl}_2, \text{ r.t., 10 h}]{\textbf{1, 2}}$ RO$_2$C—CO$_2$R + RO$_2$C / CO$_2$R

Scheme 2.

Reaction of bipymox–(*S*,*S*)–*ip* and the ruthenium complex **1** in ethanol at 60 °C gave a stable complex RuCl2[bipymox] **7**, which structure was confirmed by X–ray analysis. The bipymox–ruthenium complex **7** showed the catalytic activity to give 52% yield of **3** and **4** (72:28), 1% ee (*1R,2R*) and 13% (*1R,2S*) respectively.

Table 2. Coupling reaction of diazoacetates with Ru–pybox catalysts.[a]

N_2CHCO_2R R =	product: yield (%)	ratio maleate:fumarate
Et	83	73 : 27
i-Pr	93	78 : 22
t-Bu	88	92 : 8
l-Ment	76	83 : 17

[a] Same reaction condition in Table 1 in the absence of styrene.

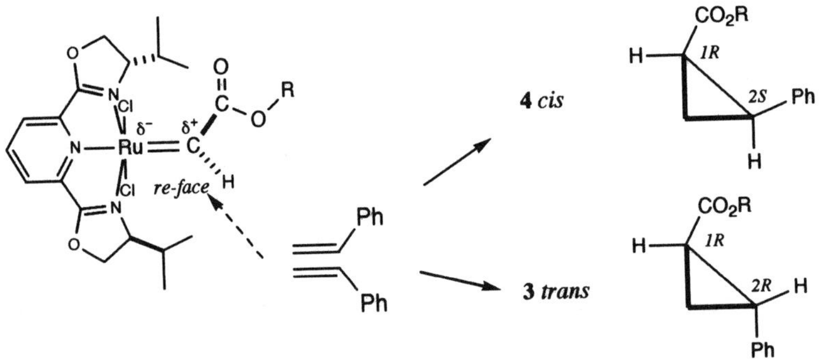

Scheme 3.

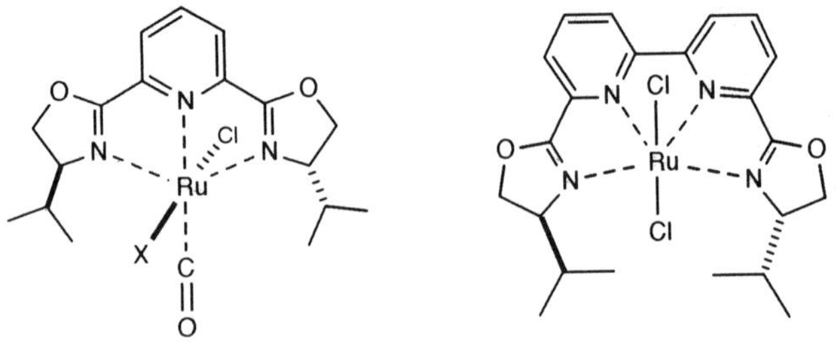

5 X = Cl
6 X = OTf

7

Scheme 4.

REFERENCES

[1] Pfaltz, A. *Acc. Chem. Res.* **1993**, *26*, 339–345.

[2] (a) Nishiyama, H., Kondo, M., Nakamura, T., Itoh, K., *Organometallics* **1991**, *100*, 500–508.

(b) Nishiyama, H.; Yamaguchi, S.; Kondo, M.; Itoh, K, *J. Org. Chem.* **1992**, *57*, 4306–4309.

[3] Nishiyama, H., Yamaguchi, S., Park, S.–B., Itoh, K., *Tetrahedron : Asymmetry* **1993**, *4*, 143–150.

[4] (a) Anciaux, A.J., Hubert, A.J., Noels, A.F., Petiniot, N., Teyssié, P. *J. Org. Chem.* **1980**, *45*, 695–702.

(b) Demonceau, A., Saive, E., de Froidmont, Y., Noels, A.F., Hubert, A.J., Chizhevsky, I.T., Lobanova, I.A., Bregadze, V.I., *Tetrahedron Lett.* **1992**, *33*, 2009–2012 : references sited therein.

(c) Maas, G., Werle, T., Alt, M., Mayer, D., *Tetrahedron* **1993**, *49*, 881–888.

[5] Bennett, M.A., Smith, A.K., J. Chem. Soc., Dalton Trans., 1974, 233–241.

[6] (a) Smith, D.A., Reynolds, D.N., Woo, L.K., *J. Am. Chem. Soc.* **1993**, *115*, 2511–2513.

(b) Woo, L.K., Smith, D.A., *Organometallics* **1992**, *11*, 2344–2346.

(c) Collman, J.P., Rose, E., Venburg, G.D., *J. Chem. Soc., Chem. Commun.* **1993**, 934–935.

[7] (a) Doyle, M.P., Brandes, B.D., Kazala, A.P., Poeters, R.J., Jarstfer, M.B., Watkins, L.M., Eagle, C.T., *Tetrehedron Lett.* **1990**, *31*, 6613–6616.

(b) O'Malley, S., Kodadek, T., *Organometallics* **1992**, *11*, 2299–2302.

POLYMER SUPPORTED *N*– ALKYLNOREPHEDRINES AS HIGHLY ENANTIOSELECTIVE CHIRAL CATALYSTS FOR THE ADDITION OF DIALKYLZINCS TO ALDEHYDES

Kenso Soai and Masami Watanabe
Department of Applied Chemistry, Faculty of Science
Science University of Tokyo
Shinjuku, Tokyo 162, Japan

INTRODUCTION

Chiral polymer supported chiral catalysts have attracted much attention [1]. In many cases, they can be easily recovered and recycled. However, the number of examples of their use in asymmetric carbon–carbon bond forming reaction is limited, and in most cases the enantiomeric excesses are low to moderate [2].

In order to realize a highly enantioselective polymer supported chiral catalyst, it is important to design a highly enantioselective monomer catalyst. We have designed several highly enantioselective chiral monomer catalysts such as *N,N*–dibutylnorephedrine (DBNE) [3], (*S*)–diphenyl-1–(methylpyrrolidin–2–yl)methanol (DPMPM) [4], chiral piperazine [5], chiral ammonium salt [6] and chiral phosphoramide [7] for the addition of dialkylzincs to aldehydes (scheme 1) [8].

Concerned with the heterogeneous chiral catalyst for the addition of dialkylzincs to aldehydes, we previously reported the enantioselective addition of dialkylzincs to aldehydes using *N*–alkylnorephedrines attached directly to chloromethylated polystyrene (scheme 2) [9,10]. The enantioselectivities of the addition to aromatic aldehydes are high (89% e.e.), however, the enantioselectivities of the addition to aliphatic aldehyde remains to be moderate (56% e.e.) [9]. We also reported the addition of dialkylzincs to aldehydes using chiral aminoalcohol supported on silica gel and alumina [11]. Although the e.e. is moderate, the reaction is the first example of the use of chiral catalysts supported on inorganic materials in enantioselective carbon–carbon bond forming reaction.

Therefore the design of a polymer supported chiral catalyst which is highly enantio-selective in the addition of dialkylzincs to aliphatic aldehydes is a challenging problem. DBNE is a highly enantioselective chiral catalyst for the addition of dialkylzincs not only to aromatic aldehydes but to *aliphatic* aldehydes [3]. The effect of *N*–alkyl substituents of *N,N*–dialkylnorephedrines is significant, and DBNE possessing *N*–butyl substituent gave optically active *sec*–alcohol with the highest e.e. (93% e.e.) in the enantioselective addition of dialkylzinc to *aliphatic* aldehyde. If the similar chiral environment of DBNE is realized in a polymer supported catalyst, the enantioselectivity of the addition to *aliphatic* aldehyde may be improved.

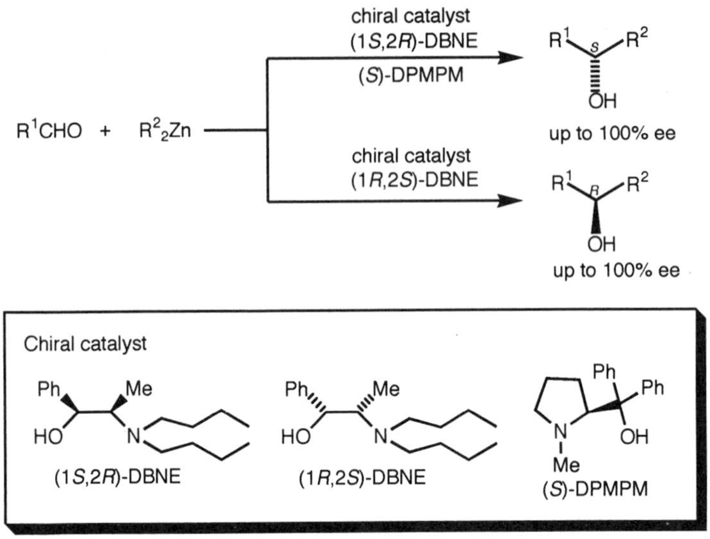

Scheme 1.

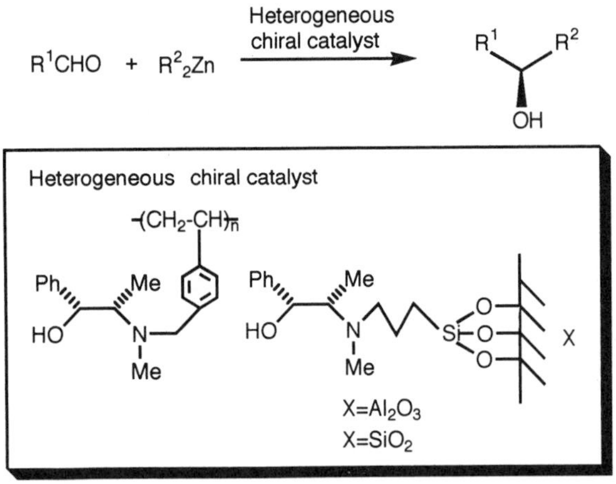

Scheme 2.

RESULTS AND DISCUSSION

In this paper, we report the design and the synthesis of new polymer supported chiral catalyst which is presently the most enantioselective among polymer supported chiral catalysts in the addition of dialkylzinc to *aliphatic* aldehyde (scheme 3).

In order to realize a chiral environment in polymer supported catalyst that be similar to that of monomer catalyst (DBNE), we designed a polymer supported chiral catalyst possessing a six–methylene spacer between the nitrogen atom of *N*–butylnorephedrine and chloromethylated polystyrene [12]. We also designed a spacer containing polymer supported chiral catalyst derived from (*S*)–proline.

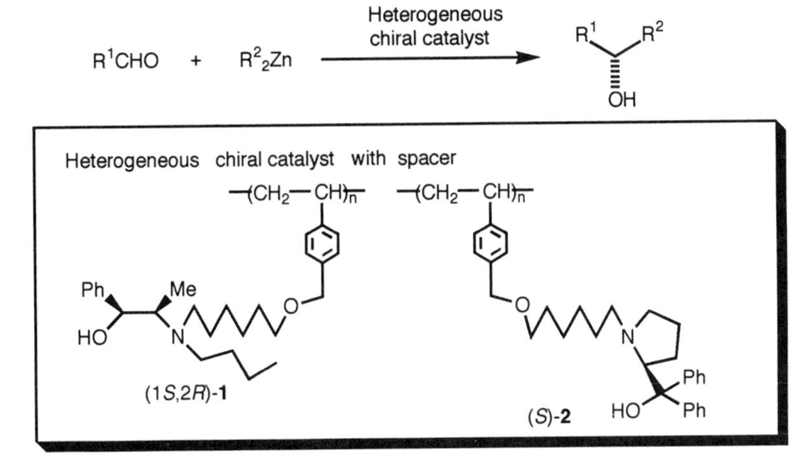

Scheme 3.

Polymer supported chiral catalysts were synthesized by the following procedures (scheme 4) : (i) 1,6–Hexanediol was converted to the corresponding polymeric mono–benzyloxyalcohol by the reaction with chloromethylated polystyrene (1% divinylbenzene; chlorine content 0.8 mmol/g; 100–200 mesh) in the presence of sodium hydride. (ii) Alcohol was converted to chloride in refluxing thionyl chloride. (iii) Treatment with sodium iodide in acetone afforded the corresponding polymeric monobenzyloxyiodide. (iv) Reaction with (1*S*, 2*R*)–*N*–butylnorephedrine [9] and (*S*)–diphenyl(pyrrolidin–2–yl)methanol [13] in the presence of potassium carbonate afforded polymer catalysts (1*S*,2*R*)–**1** and (*S*)–**2**, respectively.

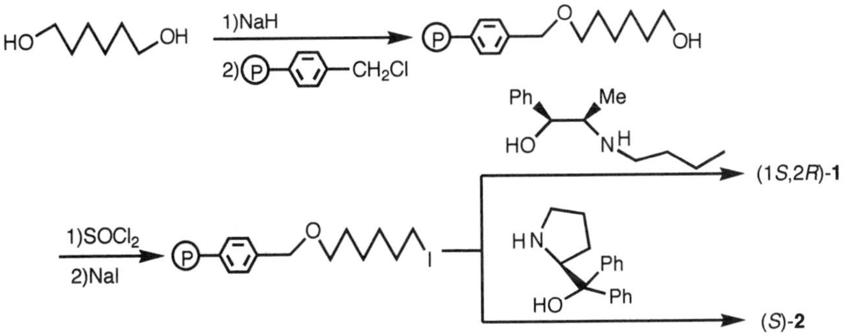

Scheme 4.

The results of the enantioselective ethylation of aromatic and aliphatic aldehydes in the presence of chiral polymer catalysts (1S, 2R)–1 and (S)–2 (content of aminoalcohol moiety to aldehyde was 20 mol%) are shown in table 1.

Table 1. Asymmetric synthesis of optically active alcohols by enantioselective addition of diethylzinc to aromatic and aliphatic aldehydes using polymer supported chiral catalyst.

			(S)–Alcohol	
Entry	Aldehyde	Chiral catalyst	Yield (%)	e.e (%)
1	PhCHO	(1S, 2R)–1	91	82
2	CH$_3$(CH$_2$)$_7$CHO	(1S, 2R)–1	75	69
3[a]	CH$_3$(CH$_2$)$_7$CHO	(1S, 2R)–1	80	71
4	trans–PhCH=CHCHO	(1S, 2R)–1	53	51
5	PhCHO	(S)–2	91	61

[a] Recycled (1S, 2R)–1 was used.

By using a polymer supported chiral catalyst (1S, 2R)–1 *with* a six–methylene spacer, the enantioselectivity of the addition to an aliphatic aldehyde improved to 71% e.e. (entry 3). The e.e. of the obtained alcohol is higher than that obtained using a polymer supported chiral catalyst [9] *without* a six–methylene spacer (64% e.e.). The enantioselectivity of the addition to aromatic aldehyde (benzaldehyde) is also high (82% e.e., entry 1). The reason of the high enantioselectivity in the special case of addition to an aliphatic aldehyde using a chiral catalyst (1S, 2R)–1 may be attributed to the methylene spacer which assures the freedom and the mobility of the reactive site of the polymer catalyst, and which realizes the similar chiral environment to that of DBNE. Chiral catalyst was easily recycled without any loss of the enantioselectivity (entries 2 and 3).

Typical experimental procedure: Aldehyde (1 mmol) was added to a suspension of polymer catalyst (0.298 g) in hexane(2 ml). The mixture was stirred for 15 min at 0 °C, then Et$_2$Zn (2.2 mmol) was added. The mixture was stirred for 1 – 8 d, then the reaction was quenched with dil. HCl. The polymer catalyst was filtered off, and the filtrate was extracted with CH$_2$Cl$_2$. The extract was dried, evaporated, and purified by silica gel TLC. Optically active *sec*–alcohol was obtained.

Concerning with the chiral catalyst (S)–2, the enantioselectivity of the addition of Et$_2$Zn to benzaldehyde is moderate (61% e.e., entry 5). Nevertheless, chiral catalyst (S)–2 is still more enantioselective than (S)–diphenyl(pyrrolidin–2–yl)methanol attached directly to polystyrene *without* methylene spacer (24% e.e.).

As described, by using polymer supported chiral catalyst with methylene spacer, optically active *sec*–alcohols with high e.e.'s are obtained from the enantioselective addition of dialkylzinc to both aromatic and aliphatic aldehydes.

ACKNOWLEDGEMENTS

This work was partly supported by The Kurata Research Grant.

REFERENCES

[1]   Review: P. Hodge, *Innovation Perspect. Solid Phase Synth. Collect. Pap., Int. Symp., 1 st,* **1989**, 273.

[2]   a) N. Kobayashi and K. Iwai, *J. Polym. Sci., Polym. Chem. Ed.,* **1980**, *18*, 923.
      b) S. Tsuboyama, *Bull. Chem. Soc. Jpn.*, **1966**, *39*, 698.
      c) For the highly enantioselective hydroformylation, G. Parrinello and J. K. Still, *J. Am. Chem. Soc.,* **1987**, *109*, 7122.

[3]   a) K. Soai, S. Yokoyama, K. Ebihara and T. Hayasaka, *J. Chem. Soc., Chem. Commun,* **1987**, 1690.
      b) K. Soai, S. Yokoyama and T. Hayasaka, *J. Org. Chem.*, **1991**, *56*, 4264.

[4]   a) K. Soai, A. Ookawa, K. Ogawa and T. Kaba, *J. Chem. Soc., Chem. Commun.*, **1987**, 467
      b) K. Soai, A. Ookawa, T. Kaba and K. Ogawa, *J. Am. Chem. Soc.,* **1987**, *109*, 7111.

[5]   a) K. Soai, S. Niwa, Y. Yamada and H. Inoue, *Tetrahedron Lett.,* **1987**, *28*, 4841.
      b) S. Niwa and K. Soai, *J. Chem. Soc., Perkin Trans. 1,* **1991**, 2717.

[6]   K. Soai and M. Watanabe, *J. Chem. Soc., Chem. Commun.*, **1990**, 43.

[7]   K. Soai, Y. Hirose and Y. Ohno, *Tetrahedron: Asymmetry,* **1993**, *4*, 1473.

[8]   Review: K. Soai and S. Niwa, *Chem. Rev.* , **1992**, *92*, 833.

[9]   a) K. Soai, S. Niwa and M. Watanabe, *J. Org. Chem.* , **1988**, *53*, 927.
      b) K. Soai, S. Niwa and M. Watanabe, *J. Chem. Soc., Perkin Trans. 1* , **1989**, 109.

[10] a) For related reactions, S. Itsuno, Y. Sakurai, K. Ito, T. Maruyama, S. Nakahama and J. M. J. Fréchet, *J. Org. Chem.,* **1990**, *55*, 304.
      b) M. Watanabe, S. Araki, Y. Butsugan and M. Uemura, *Chem. Express* , **1990**, *10*, 761.

[11] K. Soai, M. Watanabe and A. Yamamoto, *J. Org. Chem.* , **1990**, *55*, 4832.

[12] Preliminary communication: K. Soai and M. Watanabe, *Tetrahedron : Asymmetry,* **1991**, *2* , 97.

[13] E. J. Corey, S. Shibata and R. K. Bakshi, *J. Org. Chem.,* **1988**, *53*, 2861.

AUTHOR INDEX

SUBJECT INDEX